IFCoLog Journal of Logics and their Applications

Volume 4, Number 5

June 2017

Disclaimer

Statements of fact and opinion in the articles in IfCoLog Journal of Logics and their Applications are those of the respective authors and contributors and not of the IfCoLog Journal of Logics and their Applications or of College Publications. Neither College Publications nor the IfCoLog Journal of Logics and their Applications make any representation, express or implied, in respect of the accuracy of the material in this journal and cannot accept any legal responsibility or liability for any errors or omissions that may be made. The reader should make his/her own evaluation as to the appropriateness or otherwise of any experimental technique described.

© Individual authors and College Publications 2017
All rights reserved.

ISBN 978-1-84890-245-9
ISSN (E) 2055-3714
ISSN (P) 2055-3706

College Publications
Scientific Director: Dov Gabbay
Managing Director: Jane Spurr

http://www.collegepublications.co.uk

Printed by Lightning Source, Milton Keynes, UK

All rights reserved. No part of this publication may be reproduced, stored in a retrieval system or transmitted in any form, or by any means, electronic, mechanical, photocopying, recording or otherwise without prior permission, in writing, from the publisher.

Editorial Board

Editors-in-Chief
Dov M. Gabbay and Jörg Siekmann

Marcello D'Agostino
Natasha Alechina
Sandra Alves
Arnon Avron
Jan Broersen
Martin Caminada
Balder ten Cate
Agata Ciabttoni
Robin Cooper
Luis Farinas del Cerro
Esther David
Didier Dubois
PM Dung
Amy Felty
David Fernandez Duque
Jan van Eijck

Melvin Fitting
Michael Gabbay
Murdoch Gabbay
Thomas F. Gordon
Wesley H. Holliday
Sara Kalvala
Shalom Lappin
Beishui Liao
David Makinson
George Metcalfe
Claudia Nalon
Valeria de Paiva
Jeff Paris
David Pearce
Brigitte Pientka
Elaine Pimentel

Henri Prade
David Pym
Ruy de Queiroz
Ram Ramanujam
Chrtian Retoré
Ulrike Sattler
Jörg Siekmann
Jane Spurr
Kaile Su
Leon van der Torre
Yde Venema
Rineke Verbrugge
Heinrich Wansing
Jef Wijsen
John Woods
Michael Wooldridge
Anna Zamansky

Scope and Submissions

This journal considers submission in all areas of pure and applied logic, including:

pure logical systems	dynamic logic
proof theory	quantum logic
constructive logic	algebraic logic
categorical logic	logic and cognition
modal and temporal logic	probabilistic logic
model theory	logic and networks
recursion theory	neuro-logical systems
type theory	complexity
nominal theory	argumentation theory
nonclassical logics	logic and computation
nonmonotonic logic	logic and language
numerical and uncertainty reasoning	logic engineering
logic and AI	knowledge-based systems
foundations of logic programming	automated reasoning
belief revision	knowledge representation
systems of knowledge and belief	logic in hardware and VLSI
logics and semantics of programming	natural language
specification and verification	concurrent computation
agent theory	planning
databases	

This journal will also consider papers on the application of logic in other subject areas: philosophy, cognitive science, physics etc. provided they have some formal content.

Submissions should be sent to Jane Spurr (jane.spurr@kcl.ac.uk) as a pdf file, preferably compiled in LATEX using the IFCoLog class file.

CONTENTS

ARTICLES

Algorithms in Philosophy, Informatics and Logic.
  A Position Manifesto 2017 1495
  Dov M. Gabbay and Jörg Siekmann

Automation of Mathematical Induction as part of the History of Logic . . . 1505
  J Strother Moore and Claus-Peter Wirth

Algorithms in philosophy, informatics and logic. A position manifesto 2017

Dov M. Gabbay and Jörg Siekmann

The traditional view of an algorithm A is that it is a recipe, a sequence of exact steps designed to facilitate the execution of some goal involving, say, an entity E.

This view goes back to the Greeks with such well known examples as Euclid's algorithm and can be found even earlier in the Babylonian times. It is the view that has been taken up ever since by mathematicians and more recently in computer science: myriads of such algorithms are known and recorded. General investigations tried to make these notions precise and the last century in particular turned out to be very fruitful indeed with the foundational work in mathematics and computer science.

So, for example if E is a list and we wish to order it according to some measure, we may have a rich choice of algorithms for achieving this goal. They may differ in style and efficiency. Taking this view of an algorithm A, two properties come immediately to mind.

1. It is important how time and resource efficient the algorithm is.

2. The algorithm is not part of the declarative intrinsic nature of the entity it serves. Put differently — whatever the nature of the list is — the algorithm for ordering it is external to that nature.

This point of view of the relationship between algorithms and the entities they serve is commonplace. The research communities concentrated their efforts on finding algorithms for various tasks, classifying and optimising them as well as developing general theories (of algorithms) for comparing them. But, taking a platonic view, algorithms are not full entities in the ideal platonic heaven and certainly, to whatever extent that they do reside there, they are not intrinsically related to the entities they handle.

We put forward to you the reader a different view. We claim that algorithms have existence and nature beyond the local goals of what they do and that in general algorithms governing and serving an entity form an integral part of the nature of that entity.

An example taken from some century old craftsmanship is say the delineation of an elaborated cupboard in the 17th century made by a technical draughtsman to be given to the master joiner: the expert will see the drawing probably just as any layman may, but immediately perceives and mentally visualises the algorithms (or workflow to use more modern terminology) required to build this cupboard. The declarative representation is the same, but these algorithms distinguish the master from the apprentice or the journeyman — as the second author of this manifesto learned from the century old German "Handwerkskammer and Tischlerinnung", when he started off from home as a joiners apprentice.

In particular the efficiency and "Gründlichkeit" (thoroughness) by which these algorithms are executed will show in the final products and a well trained expert will be able to spot the difference in a split second.

Starting from the general debate in the mid nineteen-seventies in artificial intelligence on procedural versus declarative representation, which found its way into many technical fields such as logic programming or object oriented programming, we take the view that an entity is given by two parts: its declarative representation, i.e. its description, and the algorithms associated with it.

Put in a form of a slogan, we say

1

Object=Properties+Algorithms.

This slogan is similar to

Matter=Elementary Particles+Binding Energy.

Furthermore, we put forward that two algorithms can have distinct declarative flavours and are not just combinatorial "local hacks" and such flavours have significance across different application areas in which they are used. In fact the intuitive human mind can recognise such flavours and see affinity between algorithms in different areas which may differ in details but are similar in flavour and approach.

An animal can be described by its properties such as colour, weight and height and its parts, but just as importantly by its pattern of behaviour that make it distinct. A manufactured object has certain properties, but more likely than not, it also computes some algorithm that makes it useful: our wristwatch computes the time mechanically and an iPod uses MP3 and other electronic computations. The first radio or the most recent integrated television and web technology all derive their most interesting features from the way they function, i.e. how they compute some algorithms implemented in software or represented directly in their hardware.

To go back to our more mundane and technical example of ordered lists, a list E is characterised not only by the nature of its ordering (by size, precedence, by importance, etc.) but also by the algorithm A serving it: how to restore the order should a new element be landed in the middle of the list or should its order be disrupted and needing to be restored. So our list E is not just "E" but $\langle E, A \rangle$. And we should perceive the same E with different ordering algorithms as "different lists".

The first author of this manifesto hit upon this idea some twenty years ago in connection with logic. He was looking at goal directed formulations of various classical and non-classical logics and noticed that different logics are obtained by slight variations in the algorithm. Thus one can associate declarative properties (logic axioms) to what seems to be purely algorithmic moves (e.g. how you do your garbage collection may determine what logic you are in). Indeed we humans have an independent perception of a proof theoretical method (e.g. tableaux, resolution, truth values, etc.) as compared with our perception of the logic itself as a declarative entity. The above prompted the first author to declare that the nature of a logic itself cannot be separated from its algorithmic presentation, thus classical logic presented as a Gentzen system is not the same logic, from the point of view of applications, as classical logic presented as a resolution system (see: D. Gabbay and N. Olivetti: *Goal Directed Proof Theory*, Kluwer, 2000).

Furthermore there may be a trade-off between algorithmic optimisation constraints on a proof procedure for a logic L, and the declarative strength of L. In some cases we get a different, weaker but known logic as a result of applying these constraints, and in many cases their effects are not known. For example the connection graph method, in the theory of resolution for classical predicate logic, is a striking example of a combination of purely declarative (resolution) rules and algorithmic control (of resolution through the connections). We know that classical logic is complete for the resolution rules. However we do not know whether there is completeness for the system where resolution is controlled by a connection graph. For more than a quarter of a century this problem has been open now and it turned out to be one of the major open problems in our field (see: J. Siekmann, G. Wrightson, *Strong Completeness of Kowalski's Connection Graph Proof Procedure*, Springer Lecture Notes on AI, vol. 2408, p. 231, 2002).

The appropriate way to look at such a system is as a mixture of a declarative component presented by the complete proof procedure and the algorithmic restrictions on it. The combination of the two is a mixed presentation of a new system posing this kind of challenge.

2

This mixed presentation is present everywhere: in applied logic, AI, linguistics, computer science and even philosophy. We give more examples:

- Consider a representation of an agent (irrespective of the exact theory used). Any agent will have beliefs, belief revision, reasoning mechanisms (such as abduction) and more. Some of these features are declarative, others are algorithmic. The agent's nature is determined by both, i.e. these algorithms as well as the declarative content build the "personality" of the agent.

- Another example is from robotics. Fiora Pirri (Dipartimento di Informatica e Sistemistica Antonio Ruberti, Universita Roma) visited us some time ago telling us she won the robot rescue competition because her robot was the only one able to recognise injured persons in a room (together with many other uninjured bodies). We were impressed by this because her method was not declarative but algorithmic: the robot views a person as a compilation of body parts according to a certain algorithm. If the candidate injured person does not 'compile' correctly from its body parts, then it is an anomaly and the robot concludes that the person is indeed injured, i.e. an object is characterised not only by its declarative properties but also by the algorithms relating to its parts and the algorithms relating to how it is to be used or interact with other objects.

- This view, i.e. the inseparability of the two aspects, the algorithmic and declarative nature is of course the essence of logic programming and the above robot example is more generally present in standard object oriented programming.

- Another example can be found in theories of ambiguity and parsing in natural language. The same string of words constructed in two different ways can mean two different things. The analysis of entities such as pronouns and quantifiers require an algorithmic approach over the syntax (as evidenced by theories like Discourse Representation Theory and Dynamic Syntax). In many cases the reference of a pronoun is not determined by the grammatical algorithms alone. To identify the appropriate linguistic interpretation we use common sense and non-monotonic reasoning on the context taking into account factors such as relevance and computational effort.

- Further well known examples are SAT-procedures and model checking: essentially they are logical, i.e. declarative in nature, but these two fields derive their importance from the outmost engineering capabilities in algorithmic design and implementation.

- The following paper by J Strother Moore and Claus Peter Wirth covers the case of automated reasoning by induction in theorem proving. This is another fine case in point: the essence is the procedure (the algorithm) by which the inductive argument is carried out — in his case by a machine, i.e. a computer.

There is also a common sense human perception of the algorithmic nature of entities. Consider for example a very strict fundamentalist religious leader C_1 and a strong opposing equally fundamentalist leader A_1. They are certainly opposing entities in their nature and opinions. Compare them with more tolerant understanding and behaviourally different colleagues C_2 and A_2. The common sense reasoner may perceive that (C_1, A_1) and (C_2, A_2) have more in common than $(C_1, C_2), (A_1, A_2)$. This means that we are more perceptive of the algorithmic behavioural aspect of the entities than their declarative aspect.

The hardcore algorithmic man may remain unconvinced by the above argument. He may challenge us asking "what is the role of complexity and efficiency" in this "declarative" view of an algorithm? Given an entity E with algorithm A associated with it, suppose we manage to make A more efficient by slightly improving upon it. So we now have (E, A') instead of (E, A). We would not want to say that these are different entities but on the other hand we may not be able to say that the change from A to A' is not significant. It may be an extremely significant improvement. So what do we have to say about this?

3

Our answer lies in the aspect of potential use and potential interaction. So the two entities (E, A) and (E, A') are *not* the same because they have different potential uses and roles to play within a larger system.

Take for example an attractive supermodel and her algorithm being her mode of behaviour in public glamorous life. Let us make a small change in her behaviour, declaring that she converted to faith and religiousness and will never look at another man, except her husband. Such a small algorithmic change may make a huge difference to her social interactive life to the extent that she becomes perceptively a different person.

Potential behaviour is central to everyday life. We make our decisions today based on potential behaviour of our human and natural environment tomorrow and so a small change in the same algorithm governing some aspect of our environment may change the potentialities involved and therefore affect our decision. So (E, A) and (E, A') may be seriously different entities if they have different potentialities on account of A' being more efficient.

So what we are eluding to here is the idea that algorithms are part of the treasure trove of human knowledge collected over millennia, just like the heritage in art, social ideas (such as democracy) and any other cultural and scientific heritage we maintain. This view includes technical algorithms such as Euclid's well known case just as much as say "First-Come- First-Served" which forms part of the British way of life and socio-cultural identity to be witnessed in any London bus queue.

One of the first general repositories of this nature was set up and developed over a time span of more than two decades (see: K. Mehlhorn, S. Näher, *LEDA, A Platform for Combinatorial and Geometric Computing*, Cambridge University Press, 1999). It is now maintained by the German Max Planck Institute and a small company. Another example is the CGAL repository (Computational Geometry Algorithms Library).

An essential aspect of any algorithm is its logical nature and this is seen as its essence by many researchers, who recognise logic as being the foundational science.

$$Algorithm = Logic + Control$$

This is a slight variation of the well known slogan of R. Kowalski and the logic programming community. However this may be, i.e. runtime behaviour may be more than just "control", the essential foundational studies of the last century in logic and recursion theory provided the clarification of the nature of algorithms we enjoy today.

The first author has an image about logic and algorithms which he repeatedly tells his students: when God created the world he sprinkled around a little bit of logics (and procedures) to act as spice and bonding for his creation. Logics and algorithms are everywhere. It is the job of the research community to figure out what was given to us.

There is experimental support for the claim that objects are perceived together with the algorithms governing them. We refer to the work of professor Giacomo Rizzolatti and his colleagues. See, for example, [3].

4

The experiment, schematically described in our own words, is more or less as follows: We show a subject, (man or a trained monkey) two cups. One with a handle suitable for drinking from, and one with a different handle, not suitable for drinking. The subject responds to the drinking cup by increased activity in two regions of the brain, one dedicated to motoric actions and one known to be dedicated to the recognition of intentions and goals. The response to the non drinking cup activates only the motoric region.

This experiment and others like it clearly show that part of the perception of the drinking cup is the algorithm associated with it which includes the drinking intention. The algorithm associated with the non-drinking cup which has a different handle, does not include the drinking intention.

To summarise, we put forward to the community to recognise that algorithms are part of the nature of entities and recommend that the communities become organised socially to adjust and support this view.

5 Example of the Sorites Paradox and its Talmudic Logic Solution

A common form of the sorites paradox presented for discussion in the literature is the following form. Let F represent the soritical predicate (e.g. is bald, or does not make a heap) and let the expression a_n (where n is a natural number) represent a subject expression in the series with regard to which F is soritical (e.g. a man with n hair(s) on his head or n grain(s) of wheat). Then the sorites proceeds by way of a series of conditionals and can be schematically represented as follows:

Conditional Sorites
Fa_1
If Fa_1 then Fa_2
If Fa_2 then Fa_3
...
If Fa_{i-1} then Fa_i
Fa_i (where i can be arbitrarily large).

Whether the argument is taken to proceed by addition or subtraction will depend on how one views the series. There are many solutions to the paradox, one of them, says that indeed if one grain of sand does not make a heap (Fa_1) and adding grains of sand one by one retain this property, then any huge number of grains of sand is not a heap, Fa_n. Say for example that a collection X of 100^{100} grains is also not a heap.

The weakness of this approach is that if we start with a very huge collection Y of grains of sand with, say $N > 100^{100}$ grains which we do consider to be a heap, then if we take out of the collection one grain we still have a heap, but if we keep taking out more and more grains, we reach a collection X of 100^{100} grains, but now it will be considered still a heap. We thus have:

(*) The question of whether X is a heap depends on how it was formed.

It is at this junction that our Editorial connects with the paradox. We say the algorithm used to construct an object is part of the object and so (*) does not talk about the same heap.

We find a similar problem in the Talmud. Imagine two bottles of wine. One bottle we keep for ourselves for religious ceremony. The second bottle is given to a priest from another religion to use in his place of worship. The priest friend decants the wine, and after use, brings it back to keep in the fridge on a shelf above our own wine, which is also in a bowl. So the top bowl of wine is not kosher (no religion x would use the wine of religion y if x is not equal to y), but the bottom bowl is kosher. Unfortunately, over several days the top non-kosher wine drips, drop by drop, slowly into the bottom bowl.

The practical question is whether the wine in the bottom bowl is kosher or not?

There are many opinions of Talmudic scholars about this question. One of these opinions of Rav Dimi's on behalf of Rabbi Yochanan says as follows:

1. The initial wine in the bowl was kosher.

2. If non-kosher wine drips into a larger quantity of kosher wine then the combined quantity is kosher. (So the actual drop converts and becomes kosher.)

3. Therefore any quantity of wine obtained in this way is kosher.

So we can end up with a huge quantity of wine, 99% of which was non-kosher, but since this 99% was dripping slowly into the initial 1% which was kosher, the entire lot is now kosher.

OK. We can now ask what happens if this 99% of non-kosher wine did not drip slowly drop by drop into the 1% of kosher wine but the entire quantity of non-kosher wine just flooded into the kosher wine in one go?

Note. The answer is that the whole lot will now be not kosher. We thus get that given a bowl of mixed quantities of wine, the question of whether the wine in the bowl is kosher or not, depends on how the wine in the bowl was assembled/created as a collection. This is similar to the problem (*) above. What is the difference between the philosophical position and the Talmudic position?

1. The philosophical position sees a candidate for a heap X and has a problem in accepting that it is both a heap and not a heap, all depending on how our philosophical positions looks at it.

2. The Talmudic position of Rav Dimi's on behalf of Rabbi Yochanan would simply ask how was X created. You answer that question, and you will be told whether it is a heap or not.

The Talmud is practical, so what if you do not know how the mixture was formed? The default position is that it is not kosher. Note that the Talmudic debate is 1500 years old.

References

[1] D. Gabbay and J. Siekmann. Algorithms in cognition, informatics and logic. A position manifesto. Expanded and corrected 22.1.2008 and again 23.1.2008. Published as editorial *Logic J. IGPL* 18(6), 763-768, 2010. doi: 10.1093/jigpal/jzq004.

[2] Shlomo and Esther David, Dov Gabbay and U Schild. Talmudic logic approach to the paradox of the heap, to appear in volume on Logic and Religion, Edited by J.-Y. Beziau.

[3] M. Lacoboni, I. Molnar-Szakacs, V. Gallese, G. Buccini, J. C. Mazziotta and G. Rizzolatti. Grasping the intentions of others with one's own mirror neuron system. *Plos Biol*, 3(3):79, 2005.

Automation of Mathematical Induction as part of the History of Logic

J Strother Moore
Dept. Computer Sci., Gates Dell C., 2317 Speedway,
The University of Texas at Austin, Austin, TX 78701
moore@cs.utexas.edu

Claus-Peter Wirth
Dept. of Math., ETH Zurich, Rämistr. 101, 8092 Zürich, Switzerland
wirth@logic.at

1 A Decisive Moment in Automated Theorem Proving

The automation of mathematical theorem proving for deductive *first-order logic* started in the 1950s, and it took about half a century to develop software systems that are sufficiently strong and general to be successfully applied outside the community of automated theorem proving.[1] Even for more restricted logic languages, such as *propositional logic* and the *purely equational fragment*, such strong systems were not achieved much earlier.[2] Moreover, automation of theorem proving for *higher-order logic* has started becoming generally useful only during the last fifteen years.[3]

In this context of tedious and unexpectedly lengthy developments, it is surprising that for the field of quantifier-free first-order *inductive* theorem proving based on recursive functions, most of the progress toward general usefulness took place within the 1970s and that usefulness was clearly demonstrated by 1986.[4]

In this article we describe how this leap took place, and sketch the further development of automated inductive theorem proving.

---

[1]The currently (i.e. in 2012) most successful first-order automated theorem prover is Vampire, cf. e.g. [Riazanov & Voronkov, 2001].

[2]A breakthrough toward industrial strength in deciding propositional validity (i.e. sentential validity) (or its dual: propositional satisfiability) (which are decidable, but NP-complete) was the SAT solver Chaff, cf. e.g. [Moskewicz &al., 2001].

The most successful automated theorem prover for purely equational logic is Waldmeister, cf. e.g. [Buch & Hillenbrand, 1996], [Hillenbrand & Löchner, 2002].

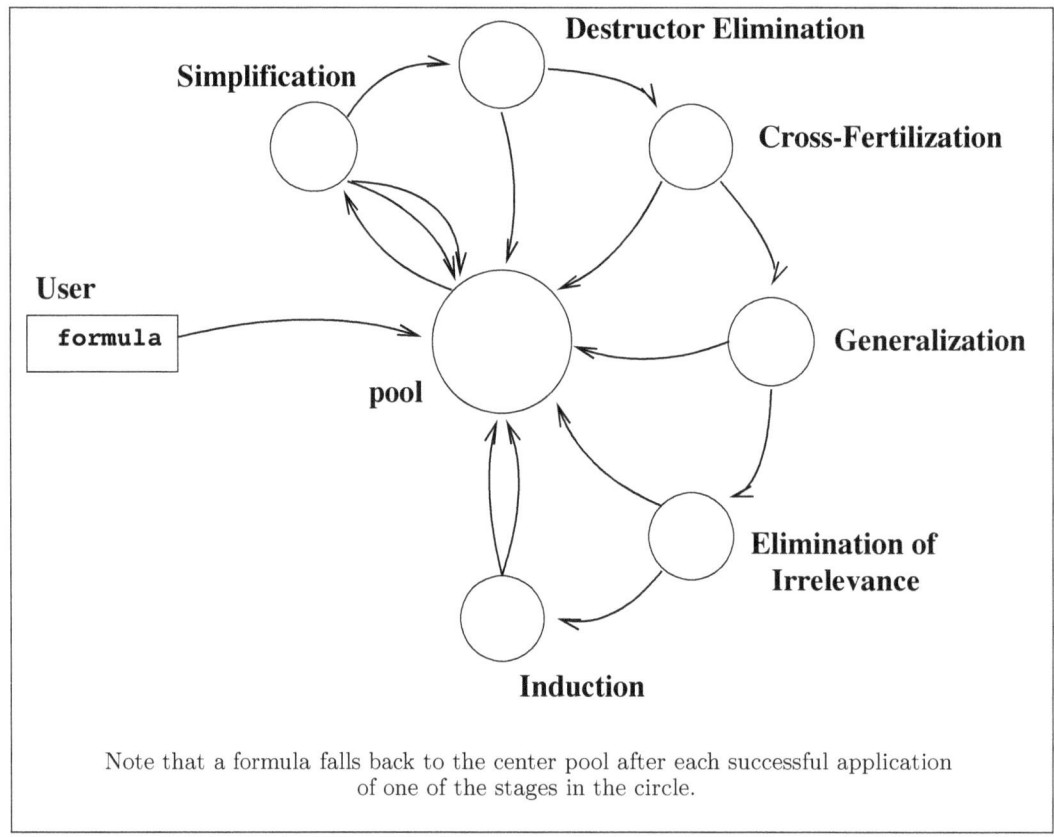

Figure 1: The Boyer–Moore Waterfall

The work on this breakthrough in the automation of inductive theorem proving was started in September 1972, by Robert S. Boyer and J Strother Moore, in Edinburgh, Scotland. Unlike the researchers in automated theorem proving until then, Boyer and Moore chose induction as the focus of their work. Most of the crucial steps and their synergetic combination in the "waterfall"[5] of their now famous theorem provers were developed in the span of a single year and implemented in their "PURE LISP THEOREM PROVER", presented at IJCAI in Stanford (CA) in August 1973,[6] and documented in Moore's PhD thesis [1973], defended in November 1973.

---

[3] Driving forces in the automation of higher-order theorem proving are the TPTP-competition-winning systems LEO-II (cf. e.g. [Benzmüller &al., 2008]) and SATALLAX (cf. e.g. [Brown, 2012]).

[4] See the last paragraph of § 6.4.

Boyer and Moore had met in August 1971, a year before the induction work started, when Boyer took up the position of a post-doctoral research fellow at the Metamathematics Unit[7] of the University of Edinburgh. Moore was at that time starting the second year of his PhD studies in "the Unit". Ironically, they were both from Texas and they had both come to Edinburgh from MIT. Boyer's PhD supervisor, W. W. Bledsoe, from The University of Texas at Austin, spent 1970–71 on sabbatical at MIT, and Boyer accompanied him and completed his PhD work there. Moore obtained his bachelor's degree at MIT (1966–70) before going to Edinburgh for his PhD.

Being "warm blooded Texans", they shared an office in the Metamathematics Unit at 9 Hope Park Square, Meadow Lane.[8] The 18th century buildings at Hope Park Square were the center of Artificial Intelligence research in Britain at a time when the promises of AI were seemingly just on the horizon. In addition to mainline work on mechanized reasoning by Rod M. Burstall, Robert A. Kowalski, Pat Hayes, Gordon Plotkin, J Strother Moore, Mike J. C. Gordon, Robert S. Boyer, Alan Bundy, and (by 1973) Robin Milner, there was work on new programming paradigms, program transformation and synthesis, natural language, machine vision, robotics, and cognitive modeling.[9]

---

[5] See Figure 1 for the Boyer–Moore waterfall. In [Bell & Thayer, 1976], credit is given to [Royce, 1970] for the probably first occurrence of "waterfall" as a term in software engineering. Boyer and Moore, however, were inspired not by this metaphor from software engineering, but again by a real waterfall, as can be clearly seen from [Boyer & Moore, 1979, p. 89]:

"A good metaphor for the organization of these heuristics is an initially dry waterfall. One pours out a clause at the top. It trickles down and is split into pieces. Some pieces evaporate as they are proved. Others are further split up and simplified. Eventually at the bottom a pool of clauses forms whose conjunction suffices to prove the original formula."

Readers who take a narrow view on the automation of inductive theorem proving might be surprised that we discuss the waterfall. It is impossible, however, to build a good inductive theorem prover without considering how to transform the induction conclusion into the hypothesis (or, alternatively, how to recognize that a legitimate induction hypothesis can dispatch a subgoal). So we take the expansive view and discuss not just the induction principle and its heuristic control, but also the waterfall architecture that is effectively an integral part of the success.

[6] Cf. [Boyer & Moore, 1973].

[7] The Metamathematics Unit of the University of Edinburgh was renamed into "Dept. of Computational Logic" in late 1971, and was absorbed into the new "Dept. of Artificial Intelligence" in Oct. 1974. It was founded and headed by Bernard Meltzer.

[8] Today's readers might have difficulty imagining the computing infrastructure in Scotland in the early 1970s.

Boyer and Moore developed their software on an ICL–4130, with 64 kByte (128 kByte in 1972) core memory (RAM). Paper tape was used for archival storage. The machine was physically located

Hope Park Square received a steady stream of distinguished visitors from around the world, including J. Alan Robinson, John McCarthy, W. W. Bledsoe, Dana S. Scott, and Marvin Minsky. An eclectic series of seminars were on offer weekly to complement the daily tea times, where all researchers gathered around a table and talked about their current problems.

Boyer and Moore initially worked together on structure sharing in resolution theorem proving. The inventor of resolution, J. Alan Robinson, created and awarded them the "1971 Programming Prize" on December 17, 1971 — half jokingly, half seriously. The document, handwritten by Robinson, actually says in part:

> "In 1971, the prize is awarded, by unanimous agreement of the Board, to Robert S. Boyer and J Strother Moore for their idea, explained in [Boyer & Moore, 1971], of representing clauses as their own genesis. The Board declared, on making the announcement of the award, that this idea is '... bloody marvelous'."

Their structure-sharing representation of derived clauses in a linear resolution system is just a stack of resolution steps. This suggests the idea of resolution being a kind of "procedure call."[10] Exploiting structure sharing, Boyer and Moore

---

in the Forrest Hill building of the University of Edinburgh, about 1 km from Hope Park Square. A rudimentary time-sharing system allowed several users at once to run lightweight applications from teletype machines at Hope Park Square.

During the day Boyer and Moore worked at Hope Park Square, with frequent trips by foot or bicycle through The Meadows to Forrest Hill to make archival paper tapes or to pick up line-printer output. During the night — when they could often have the ICL-4130 to themselves — they often worked at Boyer's home where another teletype was available.

The only high-level programming language supported was POP-2, a simple stack-based list-processing language with an ALGOL-like syntax, cf. [Burstall &al., 1971].

Programs were prepared with a primitive text editor modeled on a paper-tape editor: A disk file could be copied through a one byte buffer to an output file. By halting the copying and typing characters into or deleting characters from the buffer one could edit a file — a process that usually took several passes. Memory limitations of the ICL-4130 prohibited storing large files in memory for editing.

In their very early collaboration, Boyer and Moore solved this problem by inventing what has come to be called the "piece table", whereby an edited document is represented by a linked list of "pieces" referring to the original file which remains on disk. Their "77-editor" [Boyer &al., 1973] (written in 1971 and named for the disk track on which it resided) provided an interface like MIT's Teco, but with POP-2 as the command language. It was thus with their own editor that Boyer and Moore wrote the code for the PURE LISP THEOREM PROVER.

The 77-editor was widely used by researchers at Hope Park Square until the ICL-4130 was decommissioned. When Moore went to Xerox PARC in Palo Alto (CA) (Dec. 1973), the Boyer-Moore representation [Moore, 1981] was adopted by Charles Simonyi (*1948) for the Bravo editor on the Alto and subsequently found its way into Microsoft Word, cf. [Verma, 2005?].

implemented a declarative LISP-like programming language called "BAROQUE" [Moore, 1973], a precursor to PROLOG.[11] They then implemented a LISP interpreter in BAROQUE and began to use their resolution engine to prove simple theorems about programs in LISP. Resolution was sufficient to prove such theorems as "there is a list whose length is 3", whereas the absence of a rule for induction prevented the proofs of more interesting theorems like the associativity of list concatenation.

So, in the summer of 1972, they turned their attention to a theorem prover designed explicitly to do mathematical induction — this at a time when uniform first-order proof procedures were all the rage. The fall of 1972 found them taking turns at the blackboard, proving theorems about recursive LISP functions and articulating their reasons for each proof step. Only after several months of such proofs did they sit down together to write the POP–2 code for the PURE LISP THEOREM PROVER.

---

[9]In the early 1970s, the University of Edinburgh hosted most remarkable scientists, of which the following are relevant in our context (sources: [Meltzer, 1975], [Kowalski, 1988], etc.):

	Univ. Edinburgh (time, Dept.)	PhD (year, advisor)	life time (birth–death)
Donald Michie	(1965–1984, MI)	(1953, unknown)	(1923–2007)
Bernard Meltzer	(1965–1978, CL)	(1953, Fürth)	(1916(?)–2008)
Robin J. Popplestone	(1965–1984, MI)	(no PhD)	(1938–2004)
Rod M. Burstall	(1965–2000, MI & Dept. AI)	(1966, Dudley)	(*1934)
Robert A. Kowalski	(1967–1974, CL)	(1970, Meltzer)	(*1941)
Pat Hayes	(1967–1973, CL)	(1973, Meltzer)	(*1944)
Gordon Plotkin	(1968–today, CL & LFCS)	(1972, Burstall)	(*1946)
J Strother Moore	(1970–1973, CL)	(1973, Burstall)	(*1947)
Mike J. C. Gordon	(1970–1978, MI)	(1973, Burstall)	(*1948)
Robert S. Boyer	(1971–1973, CL)	(1971, Bledsoe)	(*1946)
Alan Bundy	(1971–today, CL)	(1971, Goodstein)	(*1947)
Robin Milner	(1973–1979, LFCS)	(no PhD)	(1934–2010)

CL = Metamathematics Unit (founded and headed by Bernard Meltzer)
      (new name from late 1971 to Oct. 1974: Dept. of Computational Logic)
      (new name from Oct. 1974: Dept. of Artificial Intelligence)
MI = Experimental Programming Unit (founded and headed by Donald Michie)
      (new name from 1966 to Oct. 1974: Dept. for Machine Intelligence and Perception)
      (new name from Oct. 1974: Machine Intelligence Unit)
LFCS = Laboratory for Foundations of Computer Science

[10]Cf. [Moore, 1973, Part I, § 6.1, pp. 68–69].

[11]For BAROQUE see [Moore, 1973, Part I, §§ 6.2 and 6.3, pp. 70–75]. For logic programming and PROLOG see [Moore, 1973, Part I, Chapter 6, pp. 68–75], [Kowalski, 1974; 1988], and [Clocksin & Mellish, 2003].

2 Method of Procedure and Presentation

The excellent survey articles [Walther, 1994a] and [Bundy, 1999] cover the engineering and research problems and current standards of the field of *explicit induction*. To cover *the history of the automation of mathematical induction*, we need a wider scope in mathematics and more historical detail. To keep this article within reasonable limits, we have to focus more narrowly on those developments and systems which are the respectively first successful and historically more important ones.

It is always hard to see the past because we look through the lens of the present. Achieving the necessary detachment from the present is especially hard for the historian of recent history because the "lens of the present" is shaped so immediately by the events being studied.

We try to mitigate this problem by avoiding the standpoint of a disciple of the leading school of explicit induction. Instead, we put the historic achievements into a broad mathematical context and a space of time from the ancient Greeks to a possible future, based on a most general approach to *recursive definition* (cf. § 5), and on *descente infinie* as a general, implementation-neutral approach to mathematical induction (cf. § 4.7). Then we can see the achievements in the field with the surprise they historically deserve — after all, until 1973 mathematical induction was considered too creative an activity to be automated.

Because the *historically most significant achievements* in the automation of inductive theorem proving manifest themselves for the first time mainly in the line of the Boyer–Moore theorem provers, we cannot avoid the confrontation of the reader with some more ephemeral forms of representation found in these software systems. In particular, we cannot avoid some small expressions in the list programming language LISP,[12] simply because the Boyer–Moore theorem provers we discuss in this article, namely the PURE LISP THEOREM PROVER, THM, NQTHM, and ACL2, all have *logics* based on a subset of LISP.

Note that we do not necessarily refer to the *implementation language* of these software systems, but to the *logic language* used both for representation of formulas and for communication with the user.

For the first system in this line of development, Boyer and Moore had a free choice, but wrote:

> "We use a subset of LISP as our language because recursive list processing functions are easy to write in LISP and because theorems can be naturally stated in LISP; furthermore, LISP has a simple syntax and

---

[12]Cf. [McCarthy &al., 1965]. Note that we use the historically correct capitalized "LISP" for general reference, but not for more recent, special dialects such as COMMON LISP.

is universal in Artificial Intelligence. We employ a LISP interpreter to
'run' our theorems and a heuristic which produces induction formulas
from information about how the interpreter fails. We combine with the
induction heuristic a set of simple rewrite rules of LISP and a heuristic
for generalizing the theorem being proved."[13]

Note that the choice of LISP was influenced by the rôle of the LISP interpreter in induction. LISP was important for another reason: Boyer and Moore were building a *computational-logic* theorem prover:

"The structure of the program is remarkably simple by artificial intelligence standards. This is primarily because the control structure is embedded in the syntax of the theorem. This means that the system does not contain two languages, the 'object language', LISP, and the 'meta-language', predicate calculus. They are identified. This mix of computation and deduction was largely inspired by the view that the two processes are actually identical. Bob Kowalski, Pat Hayes, and the nature of LISP deserve the credit for this unified view."[14]

This view was prevalent in the Metamathematics Unit by 1972. Indeed, "the Unit" was by then officially renamed the Department of Computational Logic.[7]

In general, inductive theorem proving with recursively defined functions requires a logic in which

a *method of symbolic evaluation* can be obtained from an interpretation procedure by generalizing the ground terms of computation to terms with free variables that are implicitly universally quantified.

So candidates to be considered today (besides a subset of LISP or of λ-calculus) are the typed functional programming languages ML and HASKELL,[15] which, however, were not available in 1972. LISP and ML are to be preferred to HASKELL as the logic of an inductive theorem prover because of their innermost evaluation strategy, which gives preference to the constructor terms that represent the constructor-based data types, which again establish the most interesting domains in hard- and software verification and the major elements of mathematical induction.

---

[13] Cf. [Boyer & Moore, 1973, p. 486, left column].

[14] Cf. [Moore, 1973, p. 207f.].

[15] Cf. [Hudlak &al., 1999] for HASKELL, [Paulson, 1996] for ML, which started as the meta-language for implementations of LCF (the *Logic of Computable Functions* with a single undefined element ⊥, invented by Scott [1993]) with structural induction over ⊥, 0, and s, but without original contributions to the automation of induction, cf. [Milner, 1972, p. 8], [Gordon, 2000].

Yet another candidate today would be the rewrite systems of [Wirth, 1991; 2009] and [Wirth & Gramlich, 1994a] with *constructor variables*[16] and *positive/negative-conditional equations*, designed and developed for the specification, interpretation, and symbolic evaluation of recursive functions in the context of inductive theorem proving in the domain of constructor-based data types. Neither this tailor-made theory, nor even the general theory of rewrite systems in which its development is rooted,[17] were available in 1972. And still today, the applicative subset of COMMON LISP that provides the logic language for ACL2 $(= (ACL)^2 = \underline{A}$ $\underline{C}$omputational $\underline{L}$ogic for $\underline{A}$pplicative $\underline{C}$OMMON $\underline{L}$ISP) is again to be preferred to these positive/negative-conditional rewrite systems for reasons of efficiency: The applications of ACL2 in hardware verification and testing require a performance that is still at the very limits of today's computing technology. This challenging efficiency demand requires, among other aspects, that the logic of the theorem prover is so close to its own programming language that — after certain side conditions have been checked — the theorem prover can defer the interpretation of ground terms to the analogous interpretation in its own programming language.

For most of our illustrative examples in this article, however, we will use the higher flexibility and conceptual adequacy of positive/negative-conditional rewrite systems. They are so close to standard logic that we can dispense their semantics to the reader's intuition,[18] and they can immediately serve as an intuitively clear replacement of the *Boyer–Moore machines*.[19]

Moreover, the typed (many-sorted) approach of the positive/negative-conditional equations allows the presentation of formulas in a form that is much easier to grasp for human readers than the corresponding sugar-free LISP notation with its overhead of explicit type restrictions.

Another reason for avoiding LISP notation is that we want to make it most obvious that the achievements of the Boyer–Moore theorem provers are not limited to their LISP logic.

---

[16] See § 5.4 of this article.

[17] See [Dershowitz & Jouannaud, 1990] for the theory in which the rewrite systems of [Wirth & Gramlich, 1994a], [Wirth, 1991; 2009] are rooted. One may try to argue that the paper that launched the whole field of rewrite systems, [Knuth & Bendix, 1970], was already out in 1972, but the relevant parts of rewrite theory for unconditional equations were developed only in the late 1970s and the 1980s. Especially relevant in the given context are [Huet, 1980] and [Toyama, 1988]. The rewrite theory of *positive/negative-conditional* equations, however, started to become an intensive area of research only with the burst of creativity at 1st Int. Workshop on Conditional Term Rewriting Systems (CTRS), Orsay (France), 1987; cf. [Kaplan & Jouannaud, 1988].

[18] The readers interested into the precise details are referred to [Wirth, 2009].

[19] Cf. [Boyer & Moore, 1979, p. 165f.].

For the same reason, we also prefer examples from arithmetic to examples from list theory, which might be considered to be especially supported by the LISP logic. The reader can find the famous examples from list theory in almost any other publication on the subject.[20]

In general, we tend to present the challenges and their historical solutions with the help of small intuitive examples and refer the readers interested in the very details of the implementations of the theorem provers to the published and easily accessible documents on which our description is mostly based.

Nevertheless, small LISP expression cannot completely be avoided because we have to describe the crucial parts of the historically most significant implementations and ought to show some of the advantages of LISP's untypedness.[21] The readers, however, do not have to know more about LISP than the following: A *LISP term* is either a variable symbol, or a function call of the form $(f\ t_1\ \cdots\ t_n)$, where f is a function symbol, $t_1, \ldots, t_n$ are LISP terms, and n is one of the natural numbers, which we assume to include 0.

3 Organization of this Article

This article is further organized as follows.

§§ 4 and 5 offer a self-contained reference for the readers who are not familiar with the field of mathematical induction and its automation. In § 4 we introduce the essentials of mathematical induction. In § 5 we have to become more formal regarding recursive function definitions, their consistency, termination, and induction templates and schemes.

The main part is § 6, where we present the historically most important systems in automated induction, and discuss the details of software systems for explicit induction, with a focus on the 1970s. After describing the application context in § 6.1, we present the following Boyer–Moore theorem provers: the PURE LISP THEOREM PROVER (§ 6.2), THM (§ 6.3), NQTHM (§ 6.4), and ACL2 (§ 6.5). The historically most important remaining explicit-induction systems are sketched in § 6.6.

Alternative approaches to the automation of induction that do not follow the paradigm of explicit induction are discussed in § 7.

We conclude with § 8.

---

[20]Cf. e.g. [Moore, 1973], [Boyer & Moore, 1979; 1988b; 1998], [Walther, 1994a], [Bundy, 1999], [Kaufmann &al., 2000a; 2000b].

[21]See the advantages of the untyped, type-restriction-free declaration of the shell CONS in § 6.3.

4 Mathematical Induction

In this section, we introduce mathematical induction and clarify the difference between *descente infinie* and *Noetherian, structural,* and *explicit induction*.

According to Aristotle, *induction* means to go from the special to the general, and to realize the *general* from the memorized perception of particular cases. Induction plays a major rôle in the generation of conjectures in mathematics and the natural sciences. Modern scientists design experiments to falsify a conjectured law of nature, and they accept the law as a scientific fact only after many trials have all failed to falsify it. In the tradition of Euclid, mathematicians accept a mathematical conjecture as a theorem only after a rigorous proof has been provided. According to Kant, induction is *synthetic* in the sense that it properly extends what we think to know — in opposition to *deduction*, which is *analytic* in the sense that it cannot provide us with any information not implicitly contained in the initial judgments, though we can hardly be aware of all deducible consequences.

Surprisingly, in this well-established and time-honored terminology, *mathematical induction* is not induction, but a special form of deduction for which — in the 19th century — the term "induction" was introduced and became standard in German and English mathematics.[22]

In spite of this misnomer, for the sake of brevity, the term "induction" will always refer to mathematical induction in what follows.

Although it received its current name only in the 19th century, mathematical induction has been a standard method of every working mathematician at all times. It has been conjectured[23] that Hippasus of Metapontum (ca. 550 B.C.) applied a form of mathematical induction, later named *descente infinie (ou indéfinie)* by Fermat. We find another form of induction, nowadays called *structural induction*, in a text of Plato (427–347 B.C.).[24]

In Euclid's famous "Elements" [ca. 300 B.C.], we find several applications of *descente infinie* and in a way also of structural induction.[25] Structural induction

---

[22] First in German (cf. Note 39), soon later in English (cf. [Cajori, 1918]).

[23] It is conjectured in [Fritz, 1945] that Hippasus has proved that there is no pair of natural numbers that can describe the ratio of the lengths of the sides of a pentagram and its enclosing pentagon. Note that this ratio, seen as an irrational number, is equal to the golden number, which, however, was conceptualized in entirely different terms in ancient Greek mathematics.

[24] Cf. [Acerbi, 2000].

[25] An example for *descente infinie* is Proposition 31 of Vol. VII of the Elements. Moreover, the proof in the Elements of Proposition 8 of Vol. IX seems to be sound according to mathematical standards; and so we can see it only as a proof by structural induction in a very poor linguistic and logical form. This is in accordance with [Freudenthal, 1953], but not with [Unguru, 1991] and [Acerbi, 2000]. See [Fowler, 1994] and [Wirth, 2010b, § 2.4] for further discussion.

was known to the Muslim mathematicians around the year 1000, and occurs in a Hebrew book of Levi ben Gerson (Orange and Avignon) (1288–1344).[26] Furthermore, structural induction was used by Francesco Maurolico (Messina) (1494–1575),[27] and by Blaise Pascal (1623–1662).[28] After an absence of more than one millennium (besides copying ancient proofs), *descente infinie* was reinvented by Pierre Fermat (160?–1665).[29] [30]

4.1 Well-Foundedness and Termination

A relation $<$ is *well-founded* if, for each proposition $Q(w)$ that is not constantly false, there is a $<$-minimal m among the objects for which Q holds, i.e. there is an m with $Q(m)$, for which there is no $u < m$ with $Q(u)$.

Writing "Wellf($<$)" for "$<$ is well-founded", we can formalize this *definition* as follows:

(Wellf($<$)) $\quad \forall Q. \left(\exists w.\, Q(w) \;\Rightarrow\; \exists m.\, \bigl(Q(m) \land \neg \exists u{<}m.\, Q(u)\bigr) \right)$

Let $<^+$ denote the transitive closure of $<$, and $<^*$ the reflexive closure of $<^+$.

$<$ is an (irreflexive) *ordering* if it is an irreflexive and transitive relation.

There is not much difference between a well-founded *relation* and a well-founded *ordering*:[31]

Lemma 4.1 $\quad <$ *is well-founded if and only if* $<^+$ *is a well-founded ordering.*

Closely related to the well-foundedness of a relation $<$ is the termination of its *reverse relation* written as $<^{-1}$ or $>$, and defined as $\{\,(u,v) \mid (v,u) \in <\,\}$.

---

[26] Cf. [Rabinovitch, 1970]. Also summarized in [Katz, 1998].

[27] Cf. [Bussey, 1917].

[28] Cf. [Pascal, 1954, p. 103].

[29] There is no consensus on Fermat's year of birth. Candidates are 1601, 1607 ([Barner, 2007]), and 1608. Thus, we write "160?", following [Goldstein, 2008].

[30] The best-documented example of Fermat's applications of *descente infinie* is the proof of the theorem: *The area of a rectangular triangle with positive integer side lengths is not the square of an integer;* cf. e.g. [Wirth, 2010b].

[31] Cf. Lemma 2.1 of [Wirth, 2004, § 2.1.1].

A relation $>$ is *terminating* if it has no non-terminating sequences, i.e. if there is no infinite sequence of the form $x_0 > x_1 > x_2 > x_3 \ldots$.

If $>$ has a non-terminating sequence, then this sequence, taken as a set, is a witness for the non-well-foundedness of $<$. The converse implication, however, is a weak form of the Axiom of Choice;[32] indeed, it allows us to pick a non-terminating sequence for $>$ from the set witnessing the non-well-foundedness of $<$.

So well-foundedness is slightly stronger than termination of the reverse relation, and the difference is relevant here because we cannot take the Axiom of Choice for granted in a discussion of the foundations of induction, as will be explained in § 4.3.

4.2 The Theorem of Noetherian Induction

In its modern standard meaning, the method of mathematical induction is easily seen to be a form of deduction, simply because it can be formalized as the application of the *Theorem of Noetherian Induction*:

> A proposition $P(w)$ can be shown to hold (for all w) by *Noetherian induction* over a well-founded relation $<$ as follows: *Show (for every v) that $P(v)$ follows from the assumption that $P(u)$ holds for all $u < v$.*

Again writing "Wellf($<$)" for "$<$ is well-founded", we can formalize the *Theorem of Noetherian Induction* as follows:[33]

$$(\mathsf{N}) \qquad \forall P. \left(\forall w.\ P(w) \ \Leftarrow\ \exists <. \left(\begin{array}{l} \forall v.\bigl(P(v) \Leftarrow \forall u{<}v.\ P(u)\bigr) \\ \wedge \quad \mathsf{Wellf}(<) \end{array} \right) \right)$$

The today commonly used term "Noetherian induction" is a tribute to the famous female German mathematician Emmy Noether (1882–1935). It occurs as the "Generalized principle of induction (Noetherian induction)" in [Cohn, 1965, p. 20]. Moreover, it occurs as Proposition 7 ("Principle of Noetherian Induction") in [Bourbaki, 1968a, Chapter III, § 6.5, p. 190] — a translation of the French original in its second edition [Bourbaki, 1967, § 6.5], where it occurs as Proposition 7 ("principe de récurrence nœthérienne").[34] We do not know whether "Noetherian" was used as a name of an induction principle before 1965;[35] in particular, it does not occur in the first French edition [Bourbaki, 1956] of [Bourbaki, 1967].[36]

---

[32] See [Wirth, 2004, § 2.1.2, p. 18] for the equivalence to the Principle of Dependent Choice, found in [Rubin & Rubin, 1985, p. 19], analyzed in [Howard & Rubin, 1998, p. 30, Form 43].

[33] When we write an implication $A \Rightarrow B$ in the reverse form of $B \Leftarrow A$, we do this to indicate that a proof attempt will typically focus on B and will then try to reduce the remaining open tasks to A.

4.3 An Induction Principle Stronger than Noetherian Induction?

Let us try to find a weaker replacement for the precondition of well-foundedness in Noetherian induction, in the sense that we try to replace "Wellf($<$)" in the Theorem of Noetherian Induction (N) in § 4.2 with some weaker property, which we will designate with "Weak($<, P$)" (such that $\forall P.\, \mathsf{Weak}(<, P) \Leftarrow \mathsf{Wellf}(<)$). This would result in the formula

$$(\mathsf{N}') \qquad \forall P.\left(\forall w.\, P(w) \;\Leftarrow\; \exists <.\left(\begin{array}{l}\forall v.\bigl(P(v) \Leftarrow \forall u{<}v.\, P(u)\bigr) \\ \wedge\;\; \mathsf{Weak}(<, P)\end{array}\right)\right).$$

If we assume (N'), however, we get the converse $\forall P.\, \mathsf{Weak}(<, P) \Rightarrow \mathsf{Wellf}(<)$.[37] This means that a proper weakening is possible only w.r.t. *certain* P, and the Theorem of Noetherian Induction *is the strongest among those induction principles of the form* (N') *where* $\mathsf{Weak}(<, P)$ *does not depend on* P.

C is a $<$-*chain* if $<^+$ is a total ordering on C. Let us write "$u{<}C$" for $\forall c \in C.\, u{<}c$, and "$\forall u{<}C.\, F$" as usual for $\forall u.(u{<}C \Rightarrow F)$. In [Geser, 1995], we find applications of an induction principle that roughly has the form (N') where $\mathsf{Weak}(<, P)$ is:

For every non-empty $<$-chain C [without a $<$-minimal element]:
$$\exists v \in C.\, P(v) \;\Leftarrow\; \forall u{<}C.\, P(u).$$

The resulting induction principle can be given an elegant form: If we drop the part of $\mathsf{Weak}(<, P)$ given in optional brackets [...], then we can drop the conjunction in (N') together with its first element, because $\{v\}$ is a non-empty $<$-chain.

---

[34] The peculiar French spelling "nœthérienne" imitates the German pronunciation of "Noether", where the "oe" is to be pronounced neither as a long "o" (the default, as in "Itzehoe"), nor as two separate vowels as indicated by the diaeresis in "oë", but as an umlaut, typically written in German as the ligature "ö". Neither Emmy nor her father Max Noether (1844–1921) (mathematics professor as well) used this ligature, found however in some of their official German documents.

[35] In 1967, "Noetherian Induction" was not generally used as a name for the Theorem of Noetherian Induction yet: For instance, this theorem occurs as "course-of-values induction" in [Kleene, 1952, p. 193], and as "principle of complete induction" in [Shoenfield, 1967, p. 205] (instantiated with the ordering of the natural numbers). "Complete induction", however, is a most confusing name hardly used in English. Indeed, "complete induction" is the literal translation of the German technical term "vollständige Induktion", which traditionally means structural induction (cf. Note 39) — and these two kinds of mathematical induction are different from each other.

[36] Indeed, the main text of § 6.5 in the 1st edition [Bourbaki, 1956] ends (on Page 98) three lines before the text of Proposition 7 begins in the 2nd edition [Bourbaki, 1967] (on Page 76 of § 6.5).

[37] *Proof.* Let $<\!\upharpoonright_A$ denote the range restriction of $<$ to A (i.e. $u<\!\upharpoonright_A v$ if and only if $u < v \in A$). Let us take $P(w)$ to be $\mathsf{Wellf}(<\!\upharpoonright_{A(w)})$ for $A(w) := \{\, w' \mid w'<^* w \,\}$. Then the reverse implication follows from (N') because $P(v) \Leftarrow \forall u{<}v.\, P(u)$ holds for any v,[38] and $\forall w.\, P(w)$ implies $\mathsf{Wellf}(<)$.

Then the following equivalent is obtained by switching from proposition P to its class of counterexamples Q: "If, for every non-empty $<$-chain $C \subseteq Q$, there is a $u \in Q$ with $u<C$, then $Q=\emptyset$." Under the assumption that Q is a set, this is an equivalent of the Axiom of Choice (cf. [Geser, 1995], [Rubin & Rubin, 1985]).

This means that the axiomatic status of induction principles ranges from the Theorem of Noetherian Induction up to the Axiom of Choice. If we took the Axiom of Choice for granted, this difference in status between a theorem and an axiom would collapse and our discussion of the axiomatic status of mathematical induction would degenerate. So the care with which we distinguished termination of the reverse relation from well-foundedness in § 4.1 is justified.

4.4 The Natural Numbers

The field of application of mathematical induction most familiar in mathematics is the domain of the natural numbers 0, 1, 2, ... Let us formalize the natural numbers with the help of two constructor function symbols, namely one for the constant zero and one for the direct successor of a natural number:

$$0 : \mathsf{nat}$$
$$\mathsf{s} : \mathsf{nat} \to \mathsf{nat}$$

Moreover, let us assume in this article that the variables x, y always range over the natural numbers, and that free variables in formulas are implicitly universally quantified (as is standard in mathematics), such that, for example, a formula with the free variable x can be seen as having the implicit outermost quantifier $\forall x : \mathsf{nat}$.

After the definition (Wellf($<$)) and the theorem (N), let us now consider some standard *axioms* for specifying the natural numbers, namely that a natural number is either zero or a direct successor of another natural number (nat1), that zero is not a successor (nat2), that the successor function is injective (nat3), and that the so-called *Axiom of Structural Induction over* 0 *and* s holds; formally:

---

[38] *Proof.* To show $P(v)$, it suffices to find, for an arbitrary, not constantly false proposition Q, an m with $Q(m)$, for which, in case of $m \in A(v)$, there is no $m'<m$ with $Q(m')$.

If we have $Q(m)$ for some m with $m \notin A(v)$, then we are done.

If we have $Q(u')$ for some $u < v$ and some $u' \in A(u)$, then, for $Q'(u'')$ being the conjunction of $Q(u'')$ and $u'' \in A(u)$, there is (because of the assumed $P(u)$) an m with $Q'(m)$, for which there is no $m'<m$ with $Q'(m')$. Then we have $Q(m)$. If there were an $m'<m$ with $Q(m')$, then we would have $Q'(m')$. Thus, there cannot be such an m', and so m satisfies our requirements.

Otherwise, if none of these two cases is given, Q can only hold for v. As Q is not constantly false, we get $Q(v)$ and then $v \not< v$ (because otherwise the second case is given for $u := v$ and $u' := v$). Then $m := v$ satisfies our requirements.

(nat1) $x = 0 \ \lor \ \exists y. \ (\ x = \mathsf{s}(y) \ )$
(nat2) $\mathsf{s}(x) \neq 0$
(nat3) $\mathsf{s}(x) = \mathsf{s}(y) \ \Rightarrow \ x = y$
(S) $\forall P. \ \Big(\ \forall x. \ P(x) \ \Leftarrow \ P(0) \ \land \ \forall y. \ (\ P(\mathsf{s}(y)) \Leftarrow P(y) \ ) \ \Big)$

Richard Dedekind (1831–1916) proved the Axiom of Structural Induction (S) for his model of the natural numbers in [Dedekind, 1888], where he states that the proof method resulting from the application of this axiom is known under the name "vollständige Induktion".[39]

Now we can go on by defining — in two equivalent[40] ways — the destructor function $\mathsf{p} : \mathsf{nat} \to \mathsf{nat}$, returning the predecessor of a positive natural number:

(p1) $\mathsf{p}(\mathsf{s}(x)) = x$
(p1') $\mathsf{p}(x') = x \ \Leftarrow \ x' = \mathsf{s}(x)$

The definition via (p1) is in *constructor style*, where constructor terms may occur on the left-hand side of the positive/negative-conditional equation as arguments of the function being defined. The alternative definition via (p1') is in *destructor style*, where only variables may occur as arguments on the left-hand side.

For both definition styles, the term on the left-hand side must be *linear* (i.e. all its variable occurrences must be distinct variables) and have the function symbol to be defined as the top symbol.

Let us define some recursive functions over the natural numbers, such as addition and multiplication $+, * : \mathsf{nat}, \mathsf{nat} \to \mathsf{nat}$, the irreflexive ordering of the natural numbers $\mathsf{lessp} : \mathsf{nat}, \mathsf{nat} \to \mathsf{bool}$ (see § 4.5.1 for the data type bool of Boolean values), and the Ackermann function $\mathsf{ack} : \mathsf{nat}, \mathsf{nat} \to \mathsf{nat}$:[41]

---

[39]"Vollständige Induktion" (literally: "complete induction") is a term of Aristotelian logic ("inductio completa" in [Wolff, 1740, Part I, § 478, p. 369]) and denotes a complete case analysis, cf. [Lambert, 1764, Dianoiologie, § 287; Alethiologie, § 190]. Its misuse as a designation also for mathematical induction originates in [Fries, 1822, p. 46f.] and was perpetuated by Dedekind [1888]. In the 1920s, "*das* Axiom der vollständige Induktion" ("*the* axiom of...") typically referred to Peano's axiom of structural induction (following Fries and Dedekind), cf. [Hilbert, 1926, p.117] ([Heijenoort, 1971, p.383]); the general term "vollständige Induktion", however, was not restricted to *structural* induction, cf. e.g. [Bernays, 1928, p.92] ([Heijenoort, 1971, p.489]). In English mathematics, however, "complete induction" particularly refers to a third notion, the Theorem of Noetherian Induction, cf. Note 35. Therefore, the translation of "vollständige Induktion" is "mathematical induction" throughout our text, throughout the famous source book [Heijenoort, 1971] (including complete and commented translations of [Hilbert, 1905; 1926; 1928]), and in [Eisenreich & Sube, 1982].

[40]For the equivalence transformation between constructor and destructor style see Example 6.5 in § 6.3.2.

(+1)	$0 + y = y$		(∗1)	$0 * y = 0$
(+2)	$\mathsf{s}(x) + y = \mathsf{s}(x+y)$		(∗2)	$\mathsf{s}(x) * y = y + (x * y)$

(lessp1) $\mathsf{lessp}(x, 0) = \mathsf{false}$
(lessp2) $\mathsf{lessp}(0, \mathsf{s}(y)) = \mathsf{true}$
(lessp3) $\mathsf{lessp}(\mathsf{s}(x), \mathsf{s}(y)) = \mathsf{lessp}(x, y)$

(ack1) $\mathsf{ack}(0, y) = \mathsf{s}(y)$
(ack2) $\mathsf{ack}(\mathsf{s}(x), 0) = \mathsf{ack}(x, \mathsf{s}(0))$
(ack3) $\mathsf{ack}(\mathsf{s}(x), \mathsf{s}(y)) = \mathsf{ack}(x, \mathsf{ack}(\mathsf{s}(x), y))$

The relation from a natural number to its direct successor can be formalized by the binary relation $\lambda x, y.\ (\mathsf{s}(x) = y)$. Then $\mathsf{Wellf}(\lambda x, y.\ (\mathsf{s}(x) = y))$ states the well-foundedness of this relation, which means according to Lemma 4.1 that its transitive closure — i.e. the irreflexive ordering of the natural numbers — is a well-founded ordering; so, in particular, we have $\mathsf{Wellf}(\lambda x, y.\ (\mathsf{lessp}(x, y) = \mathsf{true}))$.

Now the natural numbers can be specified up to isomorphism either by[42]

- (nat2), (nat3), and (S) — following Guiseppe Peano (1858–1932),

or else by

- (nat1) and $\mathsf{Wellf}(\lambda x, y.\ (\mathsf{s}(x) = y))$ — following Mario Pieri (1860–1913).[43]

---

[41] Rózsa Péter (1905–1977) (a female mathematician from Budapest of Jewish parentage) published [Péter, 1951, § 9(2)] a simplified version of the first recursive, but not primitive recursive function developed by Wilhelm Ackermann (1896–1962) [Ackermann, 1928]. It is actually Péter's version that is simply called "the Ackermann function" today. A very similar version already occurs in [Hilbert & Bernays, 1934, p. 332] ([Hilbert & Bernays, 1968, p. 337]) as the function ψ, for which credit is given to Rózsa Péter and to [Péter, 1932; 1935].

[42] Cf. [Wirth, 2004, § 1.1.2].

[43] Pieri [1908] stated these axioms informally and showed their equivalence to the version of the Peano axioms [Peano, 1889] given in [Padoa, 1913]. For a discussion and an English translation see [Marchisotto & Smith, 2007]. Pieri [1908] has also a version where, instead of the symbol 0, there is only the statement that there is a natural number, and where (nat1) is replaced with the weaker statement that there is at most one s-minimal element:
$$\neg \exists y_0.\ (x_0 = \mathsf{s}(y_0)) \land \neg \exists y_1.\ (x_1 = \mathsf{s}(y_1)) \Rightarrow x_0 = x_1.$$
That non-standard natural numbers cannot exist in Pieri's specification is easily shown as follows: For every natural number x we can form the set of all elements that can be reached from x by the reverse of the successor relation; by well-foundedness of s, this set contains the unique s-minimal element (0); thus, we have $x = \mathsf{s}^n(0)$ for some standard meta-level natural number n.

Immediate consequences of the axiom (nat1) and the definition (p1) are the lemma (s1) and its flattened[44] version (s1′):

(s1) $\quad \mathsf{s}(\mathsf{p}(x')) = x' \;\Leftarrow\; x' \neq 0$

(s1′) $\quad\quad \mathsf{s}(x) = x' \;\Leftarrow\; x' \neq 0 \;\wedge\; x = \mathsf{p}(x')$

Moreover, on the basis of the given axioms we can most easily show

(lessp4) $\quad \mathsf{lessp}(x, \mathsf{s}(x)) \quad\quad = \mathsf{true}$

(lessp5) $\quad \mathsf{lessp}(x, \mathsf{s}(x+y)) = \mathsf{true}$

by *structural induction on* x, i.e. by taking the predicate variable P in the Axiom of Structural Induction (S) to be $\lambda x.\,(\mathsf{lessp}(x,\mathsf{s}(x)) = \mathsf{true})$ in case of (lessp4), and $\lambda x.\,\forall y.\,(\mathsf{lessp}(x,\mathsf{s}(x+y)) = \mathsf{true})$ in case of (lessp5).

To show the necessity of doing induction on several variables in parallel, we will present[45] the more complicated proof of the strengthened transitivity of the irreflexive ordering of the natural numbers, i.e. of

(lessp7) $\quad \mathsf{lessp}(\mathsf{s}(x), z) = \mathsf{true} \;\Leftarrow\; \mathsf{lessp}(x,y) = \mathsf{true} \;\wedge\; \mathsf{lessp}(y,z) = \mathsf{true}$

We will also prove the commutativity lemma $(+3)$[46] and the simple lemma (ack4) about the Ackermann function:[47]

$(+3) \quad x+y = y+x,$

(ack4) $\quad \mathsf{lessp}(y, \mathsf{ack}(x,y)) = \mathsf{true}$

4.5 Standard Data Types

As we are interested in the verification of hardware and software, more important for us than natural numbers are the standard data types of higher-level programming languages, such as lists, arrays, and records.

To clarify the inductive character of data types defined by constructors, and to show the additional complications arising from constructors with no or more than one argument, let us present the data types bool (of Boolean values) and list(nat) (of lists over natural numbers), which we also need for some of our further examples.

---

[44] *Flattening* is an equivalence transformation that replaces a subterm (here: $\mathsf{p}(x')$) with a fresh variable (here: x) and adds a condition that equates the variable with the subterm.

[45] We will prove (lessp7) twice: once in Example 4.3 in § 4.7, and again in Example 6.2 in § 6.2.6.

[46] We will prove $(+3)$ twice: once in Example 4.2 in § 4.7, and again in Example 4.4 in § 4.8.1.

[47] We will prove (ack4) in Example 4.5 in § 4.9.

4.5.1 Boolean Values

A special case is the data type **bool** of the Boolean values given by the two constructors **true, false** : **bool** without any arguments, for which we get only the following two axioms by analogy to the axioms for the natural numbers.

(bool1) $\quad b = \text{true} \ \lor \ b = \text{false}$

(bool2) $\quad \text{true} \neq \text{false}$

We globally declare the variable b : **bool**; so b will always range over the Boolean values.

Note that the analogy of the axioms of Boolean values to the axioms of the natural numbers (cf. § 4.4) is not perfect: An axiom (bool3) analogous to (nat3) cannot exist because there are no constructors for **bool** that take arguments. Moreover, an axiom analogous to (S) is superfluous because it is implied by (bool1).

Furthermore, let us define the Boolean function **and** : **bool, bool** → **bool** :

(and1) $\quad \text{and}(\text{false}, b) \ = \ \text{false}$
(and2) $\quad \text{and}(b, \text{false}) \ = \ \text{false}$
(and3) $\quad \text{and}(\text{true}, \text{true}) \ = \ \text{true}$

4.5.2 Lists over Natural Numbers

Let us now formalize the data type of the (finite) lists over natural numbers with the help of the following two constructors: the constant symbol
$$\text{nil} : \text{list}(\text{nat})$$
for the empty list, and the function symbol
$$\text{cons} : \text{nat}, \text{list}(\text{nat}) \to \text{list}(\text{nat}),$$
which takes a natural number and a list of natural numbers, and returns the list where the number has been added to the input list as the new first element.

We globally declare the variables k, l : **list(nat)**.

By analogy to natural numbers, the axioms of this data type are the following:

(list(nat)1) $\quad l = \text{nil} \ \lor \ \exists y, k. \ (\ l = \text{cons}(y, k) \ )$

(list(nat)2) $\quad \text{cons}(x, l) \neq \text{nil}$

(list(nat)3_1) $\quad \text{cons}(x, l) = \text{cons}(y, k) \ \Rightarrow \ x = y$
(list(nat)3_2) $\quad \text{cons}(x, l) = \text{cons}(y, k) \ \Rightarrow \ l = k$

(list(nat)S) $\quad \forall P. \ \big(\forall l. \ P(l) \ \Leftarrow \ \big(P(\text{nil}) \ \land \ \forall x, k. \ \big(P(\text{cons}(x, k)) \Leftarrow P(k)\big)\big)\big)$

Moreover, let us define the recursive functions length, count : list(nat) → nat, returning the length and the size of a list:

(length1)   length(nil)   = 0
(length2)   length(cons(x, l)) = s(length(l))
(count1)   count(nil)   = 0
(count2)   count(cons(x, l)) = s(x + count(l))

Note that, just as for the Boolean values, the analogy of the axioms of lists to the axioms of the natural numbers is not perfect:

1. There is an additional axiom (list(nat)3_1), which has no analogue among the axioms of the natural numbers.

2. None of the axioms (list(nat)3_1) and (list(nat)3_2) is implied by the axiom (list(nat)1) together with the axiom
$$\text{Wellf}(\lambda l, k.\ \exists x.\ (\text{cons}(x, l) = k)),$$
which is the analogue to Pieri's second axiom for the natural numbers.[48]

3. The latter axiom is weaker than each of the two axioms
$$\text{Wellf}(\lambda l, k.\ (\text{lessp}(\text{length}(l), \text{length}(k)) = \text{true})),$$
$$\text{Wellf}(\lambda l, k.\ (\text{lessp}(\text{count}(l), \text{count}(k)) = \text{true})),$$
which state the well-foundedness of bigger[49] relations. In spite of their relative strength, the well-foundedness of these relations is already implied by the well-foundedness that Pieri used for his specification of the natural numbers.

Therefore, the lists of natural numbers can be specified up to isomorphism by a specification of the natural numbers up to isomorphism (see § 4.4), plus the axioms (list(nat)3_1) and (list(nat)3_2), plus one of the following sets of axioms:

- (list(nat)2), (list(nat)S)   — in the style of Peano,

- (list(nat)1),   Wellf($\lambda l, k.\ \exists x.\ (\text{cons}(x, l) = k)$)   — in the style of Pieri,[50]

- (list(nat)1), (length1–2)   — refining the style of Pieri.[51]

Today it is standard to take one second-order axiom (for the natural numbers) (possibly restricted to its first-order instances) as the only higher-order axiom, and to avoid further higher-order axioms in the way exemplified in the last of these three items.[52]

---

[48] See § 4.4 for Pieri's specification of the natural numbers. The axioms (list(nat)3_1) and (list(nat)3_2) are not implied because all axioms besides (list(nat)3_1) or (list(nat)3_2) are satisfied in the structure where both natural numbers and lists are isomorphic to the standard model of the natural numbers, and where lists differ only in their sizes.

Moreover, as some of the most natural functions on lists, let us define the destructors
$$\mathsf{car} : \mathsf{list}(\mathsf{nat}) \to \mathsf{nat}$$
and
$$\mathsf{cdr} : \mathsf{list}(\mathsf{nat}) \to \mathsf{list}(\mathsf{nat}),$$
both in constructor and destructor style. Furthermore, let us define the recursive member predicate
$$\mathsf{mbp} : \mathsf{nat}, \mathsf{list}(\mathsf{nat}) \to \mathsf{bool},$$
and
$$\mathsf{delfirst} : \mathsf{list}(\mathsf{nat}) \to \mathsf{list}(\mathsf{nat}),$$
a recursive function that deletes the first occurrence of a natural number in a list:

(car1) $\mathsf{car}(\mathsf{cons}(x,l)) = x$
(cdr1) $\mathsf{cdr}(\mathsf{cons}(x,l)) = l$
(car1′) $\mathsf{car}(l') = x \;\; \Leftarrow \;\; l' = \mathsf{cons}(x,l)$
(cdr1′) $\mathsf{cdr}(l') = l \;\; \Leftarrow \;\; l' = \mathsf{cons}(x,l)$

(mbp1) $\mathsf{mbp}(x, \mathsf{nil}) = \mathsf{false}$
(mbp2) $\mathsf{mbp}(x, \mathsf{cons}(y,l)) = \mathsf{true} \;\; \Leftarrow \;\; x = y$
(mbp3) $\mathsf{mbp}(x, \mathsf{cons}(y,l)) = \mathsf{mbp}(x,l) \;\; \Leftarrow \;\; x \neq y$

(delfirst1) $\mathsf{delfirst}(x, \mathsf{cons}(y,l)) = l \;\; \Leftarrow \;\; x = y$
(delfirst2) $\mathsf{delfirst}(x, \mathsf{cons}(y,l)) = \mathsf{cons}(y, \mathsf{delfirst}(x,l)) \;\; \Leftarrow \;\; x \neq y$

Immediate consequences of the axiom (list(nat)1) and the definitions (car1) and (cdr1) are the lemma (cons1) and its flattened version (cons1′):

(cons1) $\mathsf{cons}(\mathsf{car}(l'), \mathsf{cdr}(l')) = l' \;\; \Leftarrow \;\; l' \neq \mathsf{nil}$
(cons1′) $\mathsf{cons}(x, l) = l' \;\; \Leftarrow \;\; l' \neq \mathsf{nil} \;\wedge\; x = \mathsf{car}(l') \;\wedge\; l = \mathsf{cdr}(l')$

---

[49] Indeed, in case of $\mathsf{cons}(x,l) = k$, we have $\mathsf{lessp}(\mathsf{length}(l), \mathsf{length}(k)) = \mathsf{lessp}(\mathsf{length}(l), \mathsf{length}(\mathsf{cons}(x,l))) = \mathsf{lessp}(\mathsf{length}(l), \mathsf{s}(\mathsf{length}(l))) = \mathsf{true}$ because of (lessp4). Moreover, we have $\mathsf{lessp}(\mathsf{count}(l), \mathsf{count}(k)) = \mathsf{lessp}(\mathsf{count}(l), \mathsf{count}(\mathsf{cons}(x,l))) = \mathsf{lessp}(\mathsf{count}(l), \mathsf{s}(x + \mathsf{count}(l))) = \mathsf{true}$ because of (+3) and (lessp5).

[50] This option is essentially the choice of the "shell principle" of [Boyer & Moore, 1979, p.37ff.]: The one but last axiom of item (1) of the shell principle means (list(nat)2) in our formalization, and guarantees that item (6) implies $\mathsf{Wellf}(\lambda l, k.\; \exists x.\; (\mathsf{cons}(x,l) = k))$.

[51] Although (list(nat)2) follows from (length1–2) and (nat2), it should be included in this standard specification because of its frequent applications.

[52] For this avoidance, however, we have to admit the additional function length. The same can be achieved with count instead of length, which is only possible, however, for lists over element types that have a mapping into the natural numbers.

Furthermore, let us define the Boolean function
$$\mathsf{lexless} : \mathsf{list}(\mathsf{nat}), \mathsf{list}(\mathsf{nat}) \to \mathsf{bool},$$
which lexicographically compares lists according to the ordering of the natural numbers, and $\mathsf{lexlimless} : \mathsf{list}(\mathsf{nat}), \mathsf{list}(\mathsf{nat}), \mathsf{nat} \to \mathsf{bool}$, which further restricts the length of the first argument to be less than the number given as third argument:

(lexless1) $\mathsf{lexless}(l, \mathsf{nil}) = \mathsf{false}$
(lexless2) $\mathsf{lexless}(\mathsf{nil}, \mathsf{cons}(y, k)) = \mathsf{true}$
(lexless3) $\mathsf{lexless}(\mathsf{cons}(x, l), \mathsf{cons}(y, k)) = \mathsf{lexless}(l, k) \Leftarrow x = y$
(lexless4) $\mathsf{lexless}(\mathsf{cons}(x, l), \mathsf{cons}(y, k)) = \mathsf{lessp}(x, y) \Leftarrow x \neq y$
(lexlimless1) $\mathsf{lexlimless}(l, k, x) = \mathsf{and}(\mathsf{lexless}(l, k), \mathsf{lessp}(\mathsf{length}(l), x))$

Such lexicographic combinations play an important rôle in well-foundedness arguments of induction proofs, because they combine given well-founded orderings into new well-founded orderings, provided there is an upper bound for the length of the list:[53]

(lexlimless2) $\mathsf{Wellf}(\lambda l, k.\ (\mathsf{lexlimless}(l, k, x) = \mathsf{true}))$

Finally note that analogous axioms can be used to specify any other data type generated by constructors, such as pairs of natural numbers or binary trees over such pairs.

4.6 The Standard High-Level Method of Mathematical Induction

In general, the intuitive and procedural aspects of a mathematical proof method are not completely captured by its logic formalization. For actually finding and automating proofs by induction, we also need effective heuristics.

In the everyday mathematical practice of an advanced theoretical journal, the common inductive arguments are hardly ever carried out explicitly. Instead, the proof reads something like "by structural induction on n, q.e.d." or
"by (Noetherian) induction on (x, y) over $<$, q.e.d.",
expecting that the mathematically educated reader could easily expand the proof if in doubt. In contrast, difficult inductive arguments, sometimes covering several pages,[54] require considerable ingenuity and have to be carried out in the journal explicitly.

---

[53]The length limit is required because otherwise we have the following counterexample to termination: $(\mathsf{s}(0))$, $(0, \mathsf{s}(0))$, $(0, 0, \mathsf{s}(0))$, $(0, 0, 0, \mathsf{s}(0))$, ... Note that the need to compare lists of different lengths typically arises in mutual induction proofs where the induction hypotheses have a different number of free variables at measured positions. See [Wirth, 2004, § 3.2.2] for a nice example.

[54]Such difficult inductive arguments are the proofs of Hilbert's *first ε-theorem* [Hilbert & Bernays, 1970], Gentzen's *Hauptsatz* [Gentzen, 1935], and confluence theorems such as the ones in [Gramlich & Wirth, 1996], [Wirth, 2009].

In case of a proof on natural numbers, the experienced mathematician might engineer his proof roughly according to the following pattern:

> He starts with the conjecture and simplifies it by case analysis, typically based on the axiom (nat1). When he realizes that the current goal is similar to an instance of the conjecture, he applies the instantiated conjecture just like a lemma, but keeps in mind that he has actually applied an induction hypothesis. Finally, using the free variables of the conjecture, he constructs some ordering whose well-foundedness follows from the axiom $\mathsf{Wellf}(\lambda x, y.\ (\mathsf{s}(x)=y))$ and in which all instances of the conjecture applied as induction hypotheses are smaller than the original conjecture.

The hard tasks of a proof by mathematical induction are thus:

(Induction-Hypotheses Task)
> to find the numerous induction hypotheses,[55] and

(Induction-Ordering Task)
> to construct an *induction ordering* for the proof, i.e. a well-founded ordering that satisfies the ordering constraints of all these induction hypotheses in parallel.[56]

The above induction method can be formalized as an application of the Theorem of Noetherian Induction. For non-trivial proofs, mathematicians indeed prefer the axioms of Pieri's specification in combination with the Theorem of Noetherian Induction (N) to Peano's alternative with the Axiom of Structural Induction (S), because the instances for P and $<$ in (N) are often easier to find than the instances for P in (S) are.

4.7 Descente Infinie

The soundness of the induction method of § 4.6 is most easily seen when the argument is structured as a proof by contradiction, assuming a counterexample. For Fermat's historic reinvention of the method, it is thus just natural that he developed the method in terms of assumed counterexamples.[57] Here is Fermat's Method of *Descente Infinie* in modern language, very roughly speaking:

---

[55] As, e.g., in the proof of Gentzen's Hauptsatz on Cut-elimination.

[56] For instance, this was the hard part in the elimination of the ε-formulas in the proof of the 1st ε-theorem in [Hilbert & Bernays, 1970], and in the proof of the consistency of arithmetic by the ε-substitution method in [Ackermann, 1940].

[57] Cf. [Fermat, 1891ff.], [Mahoney, 1994], [Bussotti, 2006], [Wirth, 2010b].

A proposition $P(w)$ can be proved by *descente infinie* as follows: *Show that for each assumed counterexample v of P there is a smaller counterexample u of P w.r.t. a well-founded relation $<$ that is not dependent on counterexamples.*

If this method is executed successfully, we have proved $\forall w.\ P(w)$ because no counterexample can be a $<$-minimal one, and so the well-foundedness of $<$ implies that there are no counterexamples at all.

It was very hard for Fermat to obtain a positive version of his counterexample method.[58] Nowadays every logician immediately realizes that a formalization of the method of *descente infinie* is obtained from the Theorem of Noetherian Induction (N) (cf. § 4.2) simply by replacing
$$P(v) \;\Leftarrow\; \forall u{<}v.\ P(u)$$
with its contrapositive
$$\neg P(v) \;\Rightarrow\; \exists u{<}v.\ \neg P(u).$$

For the history of the *automation* of induction, however, that difference between an implication and its contrapositive is not crucial. Indeed, for this endeavor, the relevant mathematical logic was formalized during the 19th and the 20th centuries and we may confine ourselves to *classical* (i.e. two-valued) logics. What actually matters here is the heuristic task of finding proofs. Therefore — overlooking that difference — we will take *descente infinie* in the remainder of this article[59] simply as a synonym for the modern standard high-level method of mathematical induction described in § 4.6.

Let us now prove the lemmas (+3) and (lessp7) of § 4.4 (in the axiomatic context of § 4.4) by *descente infinie*, seen as the standard high-level method of mathematical induction described in § 4.6.

---

[58]Fermat reported in his letter for Christiaan Huygens (1629–1695) that he had had problems applying the Method of *Descente Infinie* to positive mathematical statements. See [Wirth, 2010b, p. 11] and the references there, in particular [Fermat, 1891ff., Vol. II, p. 432].

Moreover, a natural-language presentation via *descente infinie* (such as Fermat's representation in Latin) is often simpler than a presentation via the Theorem of Noetherian Induction, because it is easier to speak of one counterexample v and to find one smaller counterexample u, than to manage the dependences of universally quantified variables.

[59]In general, in the tradition of [Wirth, 2004], *descente infinie* is nowadays taken as a synonym for the standard high-level method of mathematical induction as described in § 4.6. This way of using the term *"descente infinie"* is found in [Brotherston & Simpson, 2007; 2011], [Voicu & Li, 2009], [Wirth, 2005a; 2010a; 2012c; 2017].

If, however, the historical perspective before the 19th century is taken, then this identification is not appropriate because a more fine-grained differentiation is required, such as found in [Bussotti, 2006], [Wirth, 2010b].

Example 4.2 (Proof of (+3) by *descente infinie*)
By application of the Theorem of Noetherian Induction (N) (cf. § 4.2) with P set to $\lambda x, y. \ (x + y = y + x)$, and the variables v, u renamed to $(x, y), (x'', y'')$, respectively, the conjectured lemma (+3) reduces to
$$\exists <. \begin{pmatrix} \forall (x, y). \ \big((x + y = y + x) \ \Leftarrow \ \forall (x'', y'') < (x, y). \ (x'' + y'' = y'' + x'')\big) \\ \wedge \ \mathsf{Wellf}(<) \end{pmatrix}.$$
Let us focus on the sub-formula $x + y = y + x$. Based on axiom (nat1) we can reduce this task to the two cases $x = 0$ and $x = \mathsf{s}(x')$ with the two goals
$$0 + y = y + 0; \qquad\qquad \mathsf{s}(x') + y = y + \mathsf{s}(x');$$
respectively. They simplify by (+1) and (+2) to
$$y = y + 0; \qquad\qquad \mathsf{s}(x' + y) = y + \mathsf{s}(x');$$
respectively. Based on axiom (nat1) we can reduce each of these goals to the two cases $y = 0$ and $y = \mathsf{s}(y')$, which leaves us with the four open goals
$$0 = 0 + 0; \qquad\qquad \mathsf{s}(x' + 0) = 0 + \mathsf{s}(x');$$
$$\mathsf{s}(y') = \mathsf{s}(y') + 0; \qquad\qquad \mathsf{s}(x' + \mathsf{s}(y')) = \mathsf{s}(y') + \mathsf{s}(x').$$
They simplify by (+1) and (+2) to
$$0 = 0; \qquad\qquad \mathsf{s}(x' + 0) = \mathsf{s}(x');$$
$$\mathsf{s}(y') = \mathsf{s}(y' + 0); \qquad\qquad \mathsf{s}(x' + \mathsf{s}(y')) = \mathsf{s}(y' + \mathsf{s}(x'));$$
respectively. Now we instantiate the induction hypothesis that is available in the context[60] given by our above formula in four different forms, namely we instantiate (x'', y'') with $(x', 0), (0, y'), (x', \mathsf{s}(y'))$, and $(\mathsf{s}(x'), y')$, respectively. Rewriting with these instances, the four goals become:
$$0 = 0; \qquad\qquad \mathsf{s}(0 + x') = \mathsf{s}(x');$$
$$\mathsf{s}(y') = \mathsf{s}(0 + y'); \qquad\qquad \mathsf{s}(\mathsf{s}(y') + x') = \mathsf{s}(\mathsf{s}(x') + y');$$
which simplify by (+1) and (+2) to
$$0 = 0; \qquad\qquad \mathsf{s}(x') = \mathsf{s}(x');$$
$$\mathsf{s}(y') = \mathsf{s}(y'); \qquad\qquad \mathsf{s}(\mathsf{s}(y' + x')) = \mathsf{s}(\mathsf{s}(x' + y')).$$
Now the first three goals follow directly from the reflexivity of equality, whereas the last goal also needs an application of our induction hypothesis: This time we have to instantiate (x'', y'') with (x', y').

Finally, we instantiate our induction ordering $<$ to the lexicographic combination of length less than 3 of the ordering of the natural numbers. If we read our pairs as two-element lists, i.e. (x'', y'') as $\mathsf{cons}(x'', \mathsf{cons}(y'', \mathsf{nil}))$, then we can set $<$ to
$$\lambda l, k. \ (\mathsf{lexlimless}(l, k, \mathsf{s}(\mathsf{s}(\mathsf{s}(0)))) = \mathsf{true}),$$
which is well-founded according to (lexlimless2) (cf. § 4.5). Then it is trivial to show that $(\mathsf{s}(x'), \mathsf{s}(y'))$ is greater than each of $(x', 0), (0, y'), (x', \mathsf{s}(y')), (\mathsf{s}(x'), y'), (x', y')$. This completes the proof of our conjecture by *descente infinie*. □

---

[60] On how this availability can be understood formally, see [Autexier, 2005].

Example 4.3 (Proof of (lessp7) by *descente infinie*)

In the previous proof in Example 4.2 we made the application of the Theorem of Noetherian Induction most explicit, and so its presentation was rather formal w.r.t. the underlying logic. Contrary to this, let us now proceed more in the vernacular of a working mathematician. Moreover, instead of $p = \text{true}$, let us just write p.

To prove the strengthened transitivity of lessp as expressed in lemma (lessp7) in the axiomatic context of § 4.4, we have to show

$$\text{lessp}(\text{s}(x), z) \;\Leftarrow\; \text{lessp}(x, y) \wedge \text{lessp}(y, z).$$

Let us reduce the last literal. To this end, we apply the axiom (nat1) once to y and once to z. Then, after reduction with (lessp1), the two base cases have an atom **false** in their conditions, abbreviating $\text{false} = \text{true}$, which is false according to (bool2), and so the base cases are true (*ex falso quodlibet*). The remaining case, where we have both $y = \text{s}(y')$ and $z = \text{s}(z')$, reduces with (lessp3) to

$$\text{lessp}(x, z') \;\Leftarrow\; \text{lessp}(x, \text{s}(y')) \wedge \text{lessp}(y', z')$$

If we apply the induction hypothesis instantiated via $\{y \mapsto y',\; z \mapsto z'\}$ to match the last literal, then we obtain the two goals

$$\text{lessp}(x, z') \;\Leftarrow\; \text{lessp}(x, \text{s}(y')) \wedge \text{lessp}(y', z') \wedge \text{lessp}(\text{s}(x), z')$$
$$\text{lessp}(x, y') \vee \text{lessp}(\text{s}(x), z') \vee \text{lessp}(x, z') \;\Leftarrow\; \text{lessp}(x, \text{s}(y')) \wedge \text{lessp}(y', z')$$

By elimination of irrelevant literals, the first goal can be reduced to the valid conjecture $\text{lessp}(x, z') \;\Leftarrow\; \text{lessp}(\text{s}(x), z')$, but we cannot obtain a lemma simpler than our initial conjecture (lessp7) by generalization and elimination of irrelevant literals from the second goal. This means that the application of the given instantiation of the induction hypothesis is useless.

Thus, instead of induction-hypothesis application, we had better apply the axiom (nat1) also to x, obtaining the cases $x = 0$ and $x = \text{s}(x')$ with the two goals — after reduction with (lessp2) and (lessp3) —

$$\text{lessp}(0, z') \;\Leftarrow\; \text{lessp}(y', z')$$
$$\text{lessp}(\text{s}(x'), z') \;\Leftarrow\; \text{lessp}(x', y') \wedge \text{lessp}(y', z'),$$

respectively. The first is trivial by (lessp1), (lessp2) after another application of the axiom (nat1) to z'. The second is just an instance of the induction hypothesis via $\{x \mapsto x',\; y \mapsto y',\; z \mapsto z'\}$. As the induction ordering we can select any of the variables of the original conjecture w.r.t. the irreflexive ordering on the natural numbers or w.r.t. the successor relation.

This completes the proof of the conjecture by *descente infinie*.

Note that we also have made clear that the given proof can only be successful with an induction hypotheses where all variables are instantiated with predecessors.

It is actually possible to show that this simple example — *ceteris paribus* — requires an induction hypothesis resulting from an instance $\{x \mapsto x'', y \mapsto y'', z \mapsto z''\}$ where, for some meta-level natural number n, we have
$$x = \mathsf{s}^{n+1}(x'') \ \wedge \ y = \mathsf{s}^{n+1}(y'') \ \wedge \ z = \mathsf{s}^{n+1}(z'').$$
□

4.8 Explicit Induction

4.8.1 From the Theorem of Noetherian Induction to Explicit Induction

To admit the realization of the standard high-level method of mathematical induction as described in § 4.6, a proof calculus should have an explicit concept of an induction hypothesis. Moreover, it would have to cope in some form with the second-order variables P and $<$ in the Theorem of Noetherian Induction (N) (cf. § 4.2), and with the second-order variable Q in the definition of well-foundedness (Wellf$(<)$) (cf. § 4.1).

Such an implementation needs special care regarding the calculus and its heuristics. For example, the theorem provers for higher-order logic with the strongest automation today[3] are yet not able to prove standard inductive theorems by just adding the Theorem of Noetherian Induction, because this theorem immediately effects an explosion of the search space. It is a main obstacle to practical usefulness of higher-order theorem provers that they are still poor in the automation of induction.

Therefore, it is probable that — on the basis of the logic calculi and the computer technology of the 1970s — Boyer and Moore would also have failed to implement induction via these human-oriented and higher-order features. Instead, they confined the concept of an induction hypothesis to the internals of single reductive inference steps — namely the applications of the so-called *induction rule* — and restricted all other inference steps to quantifier-free first-order deductive reasoning. These decisions were crucial to their success.

Described in terms of the Theorem of Noetherian Induction, this *induction rule* immediately instantiates the higher-order variables P and $<$ with first-order predicates. This is rather straightforward for the predicate variable P, which simply becomes the (properly simplified and generalized) quantifier-free first-order conjecture that is to be proved by induction, and the tuple of the free first-order variables of this conjecture takes the place of the single argument of P; cf. Example 4.4 below.

The instantiation of the higher-order variable $<$ is more difficult: Instead of a simple instantiation, the whole context of its two occurrences is transformed. For the first occurrence, namely the one in the sub-formula $\forall u{<}v.\ P(u)$, the whole sub-formula is replaced with a conjunction of instances of $P(u)$, for which u is known to be smaller than v in some lexicographic combination of given orderings that are already known to be well-founded. As a consequence, the second occurrence of $<$, i.e. the one in $\mathsf{Wellf}(<)$, simplifies to true, and so we can drop the conjunction that contains it.

At a first glance, it seems highly unlikely that there could be any framework of proof-search heuristics in which such an induction rule could succeed in implementing all applications of the Theorem of Noetherian Induction, simply because this rule has to solve the two hard tasks of an induction proof, namely the Induction-Hypotheses Task and the Induction-Ordering Task (cf. § 4.6), right at the beginning of the proof attempt, before the proof has been sufficiently developed to exhibit its structural difficulties.

Most surprisingly, but as a matter of fact, the induction rule has proved to be most successful in realizing all applications of the Theorem of Noetherian Induction required within the proof-search heuristics of the Boyer–Moore waterfall (cf. Figure 1). Essential for this success is the relatively weak quantifier-free first-order logic:

- No new symbols have to be introduced during the proof, such as the ones of quantifier elimination. Therefore, the required instances of the induction hypothesis can already be denoted when the induction rule is applied.[61]

- A general peculiarity of induction,[62] namely that the formulation of lemmas often requires the definition of new recursive functions, is aggravated by the weakness of the logic; and the user is actually required to provide further guidance for the induction rule via these new function definitions.[63]

Moreover, this success crucially depends on the possibility to generate additional lemmas that are proved by subsequent inductions, which is best shown by example.

Example 4.4 (Proof of (+3) by Explicit Induction)
Let us prove (+3) in the context of § 4.4, just as we have done already in Example 4.2 (cf. § 4.7), but now with the induction rule as the only way to apply the Theorem of Noetherian Induction.

As the conjecture is already properly simplified and concise, we instantiate $P(w)$ in the Theorem of Noetherian Induction again to the whole conjecture and reduce this conjecture by application of the Theorem of Noetherian Induction again to

$$\exists <.\ \begin{pmatrix} \forall (x,y).\ \big((x+y = y+x) \Leftarrow \forall (x'',y'') < (x,y).\ (x''+y'' = y''+x'')\big) \\ \wedge\ \mathsf{Wellf}(<) \end{pmatrix}.$$

---
[61] Cf. Note 65.
[62] See item 2 of § 4.10.
[63] Cf. § 8.

Based, roughly speaking, on a termination analysis for the function +, the heuristic of the induction rule of explicit induction suggests to instantiate $<$ to
$$\lambda(x'', y''), (x, y).\ (\mathsf{s}(x'') = x).$$
As this relation is known to be well-founded, the induction rule reduces the task based on axiom (nat1) to two goals, namely the base case
$$0 + y = y + 0;$$
and the step case
$$(\mathsf{s}(x') + y = y + \mathsf{s}(x')) \ \Leftarrow\ (x' + y = y + x').$$

This completes the application of the induction rule. Thus, instances of the induction hypothesis can no longer be applied in the further proof (except the ones that have already been added explicitly as conditions of step cases by the induction rule).

The induction rules of the Boyer–Moore theorem provers are not able to find the many instances we applied in the proof of Example 4.2. This is different for a theoretically more powerful induction rule suggested by Christoph Walther (*1950), which actually finds the proof of Example 4.2.[64] In general, however, for harder conjectures, a simulation of *descente infinie* by the induction rule of explicit induction would require an arbitrary look-ahead into the proofs, depending on the size of the structure of these proofs; thus, because the induction rule is understood to have a limited look-ahead into the proofs, such a simulation would not fall under the paradigm of explicit induction any more. Indeed, the look-ahead of induction rules into the proofs is typically not more than a single unfolding of a single occurrence of a recursive function symbol, for each such occurrence in the conjecture.

Note that the two above goals of the base and the step case can also be obtained by reducing the input conjecture with an instance of axiom (S) (cf. § 4.4), i.e. with the Axiom of Structural Induction over 0 and s. Nevertheless, the induction rule of the Boyer–Moore theorem provers is, in general, able to produce much more complicated base and step cases than those that can be obtained by reduction with the axiom (S).

Now the first goal is simplified again to $y = y + 0$, and then another application of the induction rule results in two goals that can be proved without further induction.

The second goal is simplified to
$$(\mathsf{s}(x' + y) = y + \mathsf{s}(x')) \ \Leftarrow\ (x' + y = y + x').$$

---

[64]See [Walther, 1993, p. 99f.]. On Page 100, the most interesting step case computed by Walther's induction rule is (rewritten in constructor-style):
$$\mathsf{s}(x) + \mathsf{s}(y) = \mathsf{s}(y) + \mathsf{s}(x) \ \Leftarrow\ (\ x + \mathsf{s}(y) = \mathsf{s}(y) + x \ \wedge\ \forall z.\ (z + y = y + z)\ ).$$
In practice, however, Walther's induction rule has turned out to be overall less successful when applied within a heuristic framework similar to the Boyer–Moore waterfall (cf. Figure 1).

Now we use the condition from *left* to right for rewriting only the *left*-hand side of the conclusion and then we throw away the condition completely, with the intention to obtain a stronger induction hypothesis in a subsequent induction proof. This is the famous "*cross-fertilization*" of the Boyer–Moore waterfall (cf. Figure 1). By this, the simplified second goal reduces to
$$\mathsf{s}(y+x') = y + \mathsf{s}(x').$$
Now the induction rule triggers a structural induction on y, which is successful without further induction.

All in all, although the induction rule of the Boyer–Moore theorem provers does not find the more complicated induction hypotheses of the *descente infinie* proof of Example 4.2 in § 4.7, it is well able prove our original conjecture with the help of the additional lemmas $y = y + 0$ and $\mathsf{s}(y+x') = y + \mathsf{s}(x')$.

It is crucial here that the heuristics of the Boyer–Moore waterfall discover these lemmas automatically, and that this is also typically the case in general.

From a logical viewpoint, these lemmas are redundant because they follow from the original conjecture and the definition of $+$. From a heuristic viewpoint, however, they are more useful than the original conjecture, because — oriented for rewriting from right to left — their application tends to terminate in the context of the overall simplification by symbolic evaluation, which constitutes the first stage of the waterfall. □

Although the two proofs of the very simple conjecture (+3) given in Examples 4.2 and 4.4 can only give a very rough idea on the advantage of *descente infinie* for hard induction proofs,[65] these two proofs nicely demonstrate how the induction rule of explicit induction manages to prove simple theorems very efficiently and with additional benefits for the further performance of the simplification procedure.

---

[65] For some of the advantages of *descente infinie*, see Example 6.2 in § 6.2.6, and especially the more difficult, complete formal proof of Max H. A. Newman's famous lemma in [Wirth, 2004, § 3.4], where the reverse of a well-founded relation is shown to be confluent in case of local confluence — *by induction w.r.t. this well-founded relation itself*. The induction rule of explicit induction cannot be applied here because an eager induction hypothesis generation is not possible: The required instances of the induction hypothesis contain δ-variables (or parameters, atoms) that can only be generated later during the proof by quantifier elimination.

Though confluence is the Church–Rosser property, the Newman Lemma has nothing to do with the Church–Rosser Theorem stating the confluence of the rewrite relation of $\alpha\beta$-reduction in untyped λ-calculus, which has actually been verified with a Boyer–Moore theorem prover in the first half of the 1980s by Shankar [1988] (see the last paragraph of § 6.4 and Note 176) following the short Tait/Martin-Löf proof found e.g. in [Barendregt, 2012, p. 59ff.]. Unlike the Newman Lemma, Shankar's proof proceeds by structural induction on the λ-terms, not by Noetherian induction w.r.t. the reverse of the rewrite relation; in fact, untyped λ-calculus is not terminating.

Moreover, for proving very hard theorems for which the overall waterfall heuristic fails, the user can state hints and additional lemmas with additional notions in any Boyer–Moore theorem prover, except the PURE LISP THEOREM PROVER.

4.8.2 Theoretical Viewpoint on Explicit Induction

From a theoretical viewpoint, we have to be aware of the possibility that the intended models of specifications in *explicit-induction systems* may also include non-standard models.

For the natural numbers, for instance, there may be **Z**-chains in addition to the natural numbers **N**, whereas the higher-order specifications of Peano and Pieri specify exactly the natural numbers **N** up to isomorphism.[66] This is indeed the case for the Boyer–Moore theorem provers as explained in Note 138. These **Z**-chains cannot be excluded because the inference rules realize only first-order deductive reasoning, except for the induction rule to which all applications of the Theorem of Noetherian Induction are confined and which does not use any higher-order properties, but only well-founded orderings that are defined in the first-order logic of the explicit-induction system.

4.8.3 Practical Viewpoint on Explicit Induction

Note that the application of the induction rule of explicit induction is not implemented via a reference to the Theorem of Noetherian Induction, but directly handles the following practical tasks and their heuristic decisions.

In general, the *induction stage* of the Boyer–Moore waterfall (cf. Figure 1) applies the induction rule once to its input formula, which results in a conjunction — or conjunctive set — of base and step cases to which the input conjecture reduces, i.e. whose validity implies the validity of the input conjecture.

Therefore, a working mathematician would expect that the induction rule of explicit induction solves the following two tasks:

1. Choose some of the variables in the conjecture as *induction variables*, and split the conjecture into several base and step cases, based on the induction variables' demand on which governing conditions and constructor substitutions[67] have to be added to be able to unfold — without further case analysis — some of the recursive function calls that contain the induction variables as direct arguments.

2. Eagerly generate the induction hypotheses for the step cases.

The actual realization of these tasks in the induction rule, however, is quite different from these expectations: Except the very early days of explicit induction in the PURE LISP THEOREM PROVER (cf. Example 6.1), induction variables play only a very minor rôle toward the end of the procedure (in the deletion of flawed induction schemes, cf. § 6.3.8). The focus, however, is on turning the defining equations of a recursive function symbol occurring in the conjecture immediately into whole step cases including an eagerly generated induction hypothesis tailored for each recursive occurrence; and the complementing bases case are generated only at the very end.[68]

4.9 Generalization

Contrary to merely deductive, analytic theorem proving, an input conjecture for a proof by induction is not only a task (as induction conclusion) but also a tool (as induction hypothesis) in the proof attempt. Therefore, a stronger conjecture is often easier to prove because it supplies us with a stronger induction hypothesis during the proof attempt.

Such a step from a weaker to a stronger input conjecture is called *generalization*.

Generalization is to be handled with great care because it is a sound, but *unsafe* reduction step in the sense that it may reduce a valid goal to an invalid goal, causing the proof attempt to fail; such a reduction is called *over-generalization*.

Generalization of input conjectures directly supplied by humans is rarely helpful because stating sufficiently general theorems is part of the standard mathematical training in induction. As we have seen in Example 4.4 of § 4.8.1, however, explicit induction often has to start another induction during the proof, and then the secondary, machine-generated input conjecture often requires generalization.

The two most simple syntactic generalizations are the replacement of terms with fresh universal variables and the removal of irrelevant side conditions.

In the vernacular of Boyer–Moore theorem provers, the first is simply called "generalization" and the second is called "elimination of irrelevance". They are dealt with in two consecutive stages of these names in the Boyer–Moore waterfall,

---

[66] Contrary to the **Z**-chains (which are structures similar to the integers **Z**, injectively generated from an arbitrary element via s and its inverse, where every element is greater than every standard natural number), "s-circles" cannot exist because it is possible to show by structural induction on x the two lemmas $\mathsf{lessp}(x, x) = \mathsf{false}$ and $\mathsf{lessp}(x, \mathsf{s}^{n+1}(x)) = \mathsf{true}$ for each standard meta-level natural number n.

[67] This adding of constructor substitutions refers to the application of axioms like (nat1) (cf. § 4.4), and is required whenever constructor style either is found in the recursive function definitions or is to be used for the step cases. In the PURE LISP THEOREM PROVER, only the latter is the case. In THM, none is the case.

[68] See, e.g., Example 5.5 of § 5.8.

which come right before the induction stage.

The removal of irrelevant side conditions is intuitively clear. For formulas in clausal form, it simply means to remove irrelevant literals. More interesting are the heuristics of its realization, which we discuss in § 6.3.5.

The less clear process of generalization typically proceeds by the replacement of all occurrences of a non-variable[69] term with a fresh variable.

This is especially promising for subsequent induction if the same non-variable term has multiple occurrences in the conjecture, and becomes even more promising if these occurrences are found on both sides of the same positive equation or in literals of different polarity, say in a conclusion and a condition of an implication.

To avoid *over-generalization*, subterms are to be preferred to their super-terms,[70] and one should never generalize a term of any of the following forms: a constructor term, a top level term, a term with a logical operator (such as implication or equality) as top symbol, a direct argument of a logical operator, or the first argument of a conditional (IF). Indeed, for any of these forms, the information loss by generalization is typically so high that the generalization results in an invalid conjecture.

How powerful generalization can be is best seen by the multitude of its successful automatic applications, which often surprise humans. Here is one of these:

Example 4.5 (Proof of (ack4) by Explicit Induction and Generalization)
Let us prove (ack4) in the context of § 4.4 by explicit induction. It is obvious that such a proof has to follow the definition of ack in the three cases (ack1), (ack2), (ack3), using the termination ordering of ack, which is just the lexicographic combination of its arguments. So the induction rule of explicit induction reduces the input formula (ack4) to the following goals:[71]

$\mathsf{lessp}(y, \mathsf{ack}(0, y)) = \mathsf{true};$

$\mathsf{lessp}(0, \mathsf{ack}(\mathsf{s}(x'), 0)) = \mathsf{true} \;\Leftarrow\; \mathsf{lessp}(\mathsf{s}(0), \mathsf{ack}(x', \mathsf{s}(0))) = \mathsf{true};$

$\mathsf{lessp}(\mathsf{s}(y'), \mathsf{ack}(\mathsf{s}(x'), \mathsf{s}(y'))) = \mathsf{true}$
$\Leftarrow \left(\begin{array}{l} \mathsf{lessp}(y', \mathsf{ack}(\mathsf{s}(x'), y')) = \mathsf{true} \\ \wedge \;\; \mathsf{lessp}(\mathsf{ack}(\mathsf{s}(x'), y'), \mathsf{ack}(x', \mathsf{ack}(\mathsf{s}(x'), y'))) = \mathsf{true} \end{array} \right).$

---

[69] Besides the replacement of (typically all) the occurrences of a non-variable term, there is also the possibility of replacing some — *but not all* — occurrences of a variable with a fresh variable. This is a very delicate process, but heuristics for it were discussed very early, namely in [Aubin, 1976, § 3.3].

[70] This results in a weaker conjecture and the stronger one remains available by a further generalization.

[71] See Example 5.5 of § 5.8 on how these step cases are actually found in explicit induction.

After simplifying with (ack1), (ack2), (ack3), respectively, we obtain:

$\mathsf{lessp}(y, \mathsf{s}(y)) = \mathsf{true}$;

$\mathsf{lessp}(0, \mathsf{ack}(x', \mathsf{s}(0))) = \mathsf{true} \Leftarrow \mathsf{lessp}(\mathsf{s}(0), \mathsf{ack}(x', \mathsf{s}(0))) = \mathsf{true}$;

$\mathsf{lessp}(\mathsf{s}(y'), \mathsf{ack}(x', \mathsf{ack}(\mathsf{s}(x'), y'))) = \mathsf{true}$
$$\Leftarrow \left(\begin{array}{l} \mathsf{lessp}(y', \mathsf{ack}(\mathsf{s}(x'), y')) = \mathsf{true} \\ \wedge \;\; \mathsf{lessp}(\mathsf{ack}(\mathsf{s}(x'), y'), \mathsf{ack}(x', \mathsf{ack}(\mathsf{s}(x'), y'))) = \mathsf{true} \end{array} \right).$$

Now the base case is simply an instance of our lemma (lessp4). Let us simplify the two step cases by introducing variables for their common subterms (by a partial "flattening"):

$\mathsf{lessp}(0, z) = \mathsf{true} \;\Leftarrow\; (\; \mathsf{lessp}(\mathsf{s}(0), z) = \mathsf{true} \;\wedge\; z = \mathsf{ack}(x', \mathsf{s}(0)) \;)$;

$\mathsf{lessp}(\mathsf{s}(y'), z_2) = \mathsf{true} \Leftarrow \left(\begin{array}{l} \mathsf{lessp}(y', z_1) = \mathsf{true} \;\wedge\; \mathsf{lessp}(z_1, z_2) = \mathsf{true} \\ \wedge \;\; z_1 = \mathsf{ack}(\mathsf{s}(x'), y') \;\wedge\; z_2 = \mathsf{ack}(x', z_1) \end{array} \right).$

Now the first follows from applying (nat1) to z. Before we can prove the second by another induction, however, we have to generalize it by deleting the last two literals from the condition ("elimination of irrelevance"). In fact, the result of this generalization is the lemma (lessp7) of § 4.4. □

In combination with explicit induction, generalization becomes especially powerful in the invention of new lemmas of general interest, because the step cases of explicit induction tend to have common occurrences of the same term in their conclusion and their condition. Indeed, the lemma (lessp7), which we have just discovered in Example 4.5, is one of the most useful lemmas in the theory of natural numbers.

It should be noted that all Boyer–Moore theorem provers except the PURE LISP THEOREM PROVER are able to do this whole proof completely automatically and invent the lemma (lessp7) by generalization of the second step case; and they do this even when they work with an arithmetic theory that was redefined, so that no decision procedures or other special knowledge on the natural numbers can be used by the system.

Moreover, as shown in § 3.3 of [Wirth, 2004], in a slightly richer logic, these heuristics can actually synthesize the lower bound in the first argument of lessp from the weaker input conjecture $\exists z. \; (\mathsf{lessp}(z, \mathsf{ack}(x, y)) = \mathsf{true})$, simply because lessp does not contribute to the choice of the base and step cases.

4.10 Proof-Theoretical Peculiarities of Mathematical Induction

The following two proof-theoretical peculiarities of induction compared to first-order deduction may be considered noteworthy:[72]

1. A calculus for arithmetic cannot be complete, simply because the theory of the arithmetic of natural numbers is not enumerable.[73]

2. According to Gentzen's Hauptsatz,[74] a proof of a first-order theorem can always be restricted to the "sub"-formulas of this theorem. In contrast to lemma application in a deductive proof tree, however, the application of induction hypotheses and lemmas inside an inductive reasoning cycle cannot generally be eliminated in the sense that the "sub"-formula property could be obtained.[75] As a consequence, in first-order inductive theorem proving, "creativity" cannot be restricted to finding just the proper instances, but may require the invention of new lemmas and notions.[76]

4.11 Conclusion

In this section, after briefly presenting the induction method in its rich historical context, we have offered a formalization and a first practical description. Moreover, we have explained why we can take Fermat's term *"descente infinie"* in our modern context as a synonym for the standard high-level method of mathematical induction. Finally, we have introduced explicit induction and generalization.

Noetherian induction requires domains for its well-founded orderings; and these domains are typically built-up by constructors. Therefore, the discussion of the method of induction required the introduction of some paradigmatic data types, such as natural numbers and lists.

To express the relevant notions on these data types, we need *recursion*, a method of definition, which we have often used in this section intuitively. We did not discuss its formal admissibility requirements yet. We will do so in § 5, with a focus on modes of recursion that admit an effective consistency test, including termination aspects such as induction templates and schemes.

---

[72] Note, however, that these peculiarities of induction do not make a difference to first-order deductive theorem proving *in practice*. See Notes 73 and 76.

[73] This theoretical result is given by Gödel's first incompleteness theorem [1931]. In practice, however, it does not matter whether our proof attempt fails because our theorem will not be enumerated ever, or will not be enumerated before doomsday.

[74] Cf. [Gentzen, 1935].

[75] Cf. [Kreisel, 1965].

5 Recursion, Termination, and Induction

5.1 Recursion and the Rewrite Relation on Ground Terms

Recursion is a form of programming or definition where a newly defined notion may even occur in its *definientia*. Contrary to *explicit* definitions, where we can always get rid of the new notions by reduction (i.e. by rewriting the *definienda* (*left-hand sides* of the defining equations) to the *definientia* (*right-hand sides*)), reduction with *recursive* definitions may run forever.

We have already seen some recursive function definitions in §§ 4.4 and 4.5, such as the ones of +, lessp, length, and count, where these function symbols occurred in some of the right-hand sides of the equations of their own definitions; for instance, the function symbol + occurs in the right-hand side of (+2) in § 4.4.

The steps of rewriting with recursive definitions can be formalized as a binary relation on terms, namely as the *rewrite relation* that results from reading the defining equations as reduction rules, in the sense that they allow us to replace occurrences of left-hand sides of instantiated equations with their respective right-hand sides, provided that their conditions are fulfilled.[77]

A *ground* term is a term without variables. We can restrict our considerations here to rewrite relations *on ground terms*.

5.2 Confluence

The restriction that is to be required for every recursive function definition is the *confluence*[78] of this rewrite relation on ground terms.

The confluence restriction guarantees that no distinct objects of the data types can be equated by the recursive function definitions.[79]

---

[76] In practice, however, proof search for harder theorems often requires the introduction of lemmas, functions, and relations, and it is only a matter of degree whether we have to do this for principled reasons (as in induction) or for tractability (as required in first-order deductive theorem proving, cf. [Baaz & Leitsch, 1995]).

[77] For the technical meaning of *fulfilledness* in the recursive definition of the rewrite relation see [Wirth, 2009], where it is also explained why the rewrite relation respects the straightforward purely logical, model-theoretic semantics of positive/negative-conditional equation equations, provided that the given admissibility conditions are satisfied (as is the case for all our examples).

[78] A relation $\longrightarrow$ is *confluent* (or has the "Church–Rosser property") if two sequences of steps with $\longrightarrow$, starting from the same element, can always be joined by an arbitrary number of further steps on each side; formally: $\xleftarrow{+} \circ \xrightarrow{+} \subseteq \xrightarrow{*} \circ \xleftarrow{*}$. Here $\circ$ denotes the concatenation of binary relations; for the further notation see § 4.1.

[79] As constructor terms are irreducible w.r.t. this rewrite relation, if the application of a defined function symbol rewrites to two constructor terms, they must be identical in case of confluence.

This is essential for consistency if we assume axioms such as (nat2–3) (cf. § 4.4) or (list(nat)2–3) (cf. § 4.5).

Indeed, without confluence, a definition of a recursive function could destroy the data type in the sense that the specification has no model anymore; for example, if we added $p(x) = 0$ as a further defining equation to (p1), then we would get $s(0) = p(s(s(0))) = 0$, in contradiction to the axiom (nat2) of § 4.4.

For the recursive function definitions admissible in the Boyer–Moore theorem provers, confluence results from the restrictions that there is only one (unconditional) defining equation for each new function symbol,[80] and that all variables occurring on the right-hand side of the definition also occur on the left-hand side of the defining equation.[81]

These two restrictions are an immediate consequence of the general definition style of the list-programming language LISP. More specifically, recursive functions are to be defined in all Boyer–Moore theorem provers in the more restrictive style of *applicative* LISP.[82]

Example 5.1 (A Recursive Function Definition in Applicative LISP)
Instead of our two equations (+1), (+2) for +, we find the following single equation on Page 53 of [Boyer & Moore, 1979], the standard reference for the Boyer–Moore heuristics:
```
(PLUS X Y) = (IF (ZEROP X)
                 (FIX Y)
                 (ADD1 (PLUS (SUB1 X) Y)))
```
Note that (IF x y z) is nothing but the conditional "IF x then y else z", that ZEROP is a Boolean function checking for being zero, that (FIX Y) returns Y if Y is a natural number, and that ADD1 is the successor function s.

The primary difference to (+1), (+2) is that PLUS is defined in *destructor style* instead of the *constructor style* of our equations (+1), (+2) in § 4.4. As a constructor-style definition can always be transformed into an equivalent destructor-style definition, let us do so for our definition of + via (+1), (+2).

In place of the untyped destructor SUB1, let us use the typed destructor p defined

---

[80] Cf. item (a) of the "definition principle" of [Boyer & Moore, 1979, p. 44f.]. Confluence is also discussed under the label "uniqueness" on Page 87ff. of [Moore, 1973].

[81] Cf. item (c) of the "definition principle" of [Boyer & Moore, 1979, p. 44f.].

[82] See [McCarthy &al., 1965] for the definition of LISP. The "'applicative" subset of LISP lacks side effects via global variables and the imperative commands of LISP, such as variants of PROG, SET, GO, and RETURN, as well as all functions or special forms that depend on the concrete allocation on the system heap, such as EQ, RPLACA, and RPLACD, which can be used in LISP to realize circular structures or to save space on the system heap.

by either by (p1) or by (p1′) of § 4.4, which — just as SUB1 — returns the predecessor of a positive natural number. Now our destructor-style definition of + consists of the following two positive/negative-conditional equations:

(+1′) $\quad x + y = y \qquad\qquad \Leftarrow\ x = 0$
(+2′) $\quad x + y = \mathsf{s}(\mathsf{p}(x) + y) \ \Leftarrow\ x \neq 0$

If we compare this definition of + to the one via the equations (+1), (+2), then we find that the constructors 0 and s have been removed from the left-hand sides of the defining equations; they are replaced with the destructor p on the right-hand side and with some conditions.

Now it is easy to see that (+1′), (+2′) represent the above definition of PLUS in positive/negative-conditional equations, provided that we ignore that Boyer–Moore theorem provers have no types and no typed variables. □

If we considered the recursive equation (+2) together with the alternative recursive equation (+2′), then we could rewrite $\mathsf{s}(x) + y$ on the one hand with (+2) into $\mathsf{s}(x + y)$, and, on the other hand, with (+2′) into $\mathsf{s}(\mathsf{p}(\mathsf{s}(x)) + y)$. This does not seem to be problematic, because the latter result can be rewritten to the former one by (p1).

In general, however, confluence is undecidable and criteria sufficient for confluence are extremely hard to develop. The only known decidable criterion that is sufficient for confluence of conditional equations and applies to all our example specifications, but does not require termination, is found in [Wirth, 2009].[83] It can be more easily tested than the admissibility conditions of the Boyer–Moore theorem provers and avoids divergence even in case of non-termination; the proof that it indeed guarantees confluence is very involved.

5.3 Termination and Reducibility

There are two restrictions that are additionally required for any function definition in the Boyer–Moore theorem provers, namely *termination* of the rewrite relation and *reducibility* of all ground terms that contain a defined function symbol w.r.t. the rewrite relation.

The requirement of termination should be intuitively clear; we will further discuss it in § 5.5.

Let us now discuss the requirement of reducibility.

---

[83] The effective confluence test of [Wirth, 2009] requires *binding-triviality* or *-complementary* of every critical peak, and *effective weak-quasi-normality*, i.e. that each equation in the condition must be restricted to constructor variables (cf. § 5.4), or that one of its top terms either is a constructor term or occurs as the argument of a definedness literal in the same condition.

First of all, note that it is not only so that we can check the soundness of (+1′) and (+2′) independently from each other, we can even omit one of the equations, resulting in a partial definition of the function +. Indeed, for the function p we did not specify any value for p(0); so p(0) is not reducible in the rewrite relation that results from reading the specifying equations as reduction rules.

A function defined in a Boyer–Moore theorem prover, however, must always be specified completely, in the sense that every application of such a function to (constructor) ground terms must be reducible. This reducibility immediately results from the LISP definition style, which requires all arguments of the function symbol on the left-hand side of its defining equation to be distinct variables.[84]

5.4 Constructor Variables

These restrictions of reducibility and termination of the rewrite relation are not essential; neither for the semantics of recursive function definitions with data types given by constructors,[85] nor for confluence and consistency.[86]

Note that these two restrictions imply that only *total recursive* functions[87] are admissible in the Boyer–Moore theorem provers.

As a termination restriction is not in the spirit of the LISP logic of the Boyer–Moore theorem provers, we have to ask ourselves:

Why did Boyer and Moore bring up this strong additional restriction?

The following cannot count as a satisfactory answer: When both reducibility and termination are given, then — similar to the classical case of explicitly defined notions — we can get rid of all recursively defined function symbols by rewriting, but in general only for *ground* terms.

A better potential answer is found on Page 87ff. of [Moore, 1973], where confluence of the rewrite relation is discussed and a reference to Russell's Paradox serves as an argument that confluence alone would not be sufficient for consistency. The argumentation is essentially the following: First, a Boolean function russell is recursively defined by

(russell1) $\quad$ russell(b) = false $\;\Leftarrow\;$ russell(b) = true
(russell2) $\quad$ russell(b) = true $\;\Leftarrow\;$ russell(b) = false

---

[84] Cf. item (b) of the "definition principle" of [Boyer & Moore, 1979, p. 44f.].

[85] Cf. [Wirth & Gramlich, 1994b].

[86] Cf. [Wirth, 2009].

[87] You may follow the explicit reference to [Shoenfield, 1967] as the basis for the logic of the PURE LISP THEOREM PROVER on Page 93 of [Moore, 1973].

Then it is claimed that this function definition would result in an inconsistent specification on the basis of the axioms (bool1–2) of § 4.5.

This inconsistency, however, arises only if the variable b of the axiom (bool1) can be instantiated with the term russell(b), which is actually not our intention and which we do not have to permit: If all variables we have introduced so far are *constructor variables*[88] in the sense that they can only be instantiated with terms formed from constructor function symbols (incl. constructor constants) and constructor variables, then irreducible terms such as russell(b) can denote *junk objects* different from true and false, and no inconsistency arises.[89]

Note that these constructor variables are implicitly part of the LISP semantics with its innermost evaluation strategy. For instance, in Example 5.1 of § 5.2, neither the LISP definition of PLUS nor its representation via the positive/negative-conditional equations (+1′), (+2′) is intended to be applied to a non-constructor term in the sense that X or x should be instantiated to a term that is a function call of a (partially) defined function symbol that may denote a junk object.

Moreover, there is evidence that Moore considered the variables already in 1973 as constructor variables: On Page 87 in [Moore, 1973], we find formulas on definedness and confluence, which make sense only for constructor variables; the one on definedness of the Boolean function (AND X Y) reads[90]

$$\exists Z \ (\text{IF X (IF Y T NIL) NIL}) = Z,$$

which is trivial for a general variable Z and makes sense only if Z is taken to be a constructor variable.

Finally, the way termination is established via induction templates in Boyer–Moore theorem provers and as we will describe it in § 5.5, is sound for the rewrite relation of the defining equations only if we consider the variables of these equations to be constructor variables (or if we restrict the termination result to an innermost rewriting strategy and require that all function definitions are total).

---

[88] Such *constructor variables* were formally introduced for the first time in [Wirth &al., 1993] and became an essential part of the frameworks found in [Wirth & Gramlich, 1994a; 1994b], [Kühler & Wirth, 1996; 1997], [Wirth, 1997; 2009] [Kühler, 2000], [Avenhaus &al., 2003], and [Schmidt-Samoa, 2006a; 2006b; 2006c].

[89] For the appropriate semantics see [Wirth & Gramlich, 1994b], [Kühler & Wirth, 1997].

[90] In the logic of the PURE LISP THEOREM PROVER, the special form IF is actually called "COND". This is most confusing because COND is a standard special form in LISP, different from IF. Therefore, we will ignore this peculiarity and tacitly write "IF" here and in what follows for every "COND" of the PURE LISP THEOREM PROVER.

5.5 Termination and General Induction Templates

In addition to the restricted style of recursive definition that is found in LISP and that guarantees reducibility of terms with defined function symbols and confluence as described in §§ 5.3 and 5.4, the theorem provers for explicit induction require termination of the rewrite relation that results from reading the specifying equations as reduction rules. More precisely, in all Boyer–Moore theorem provers except the PURE LISP THEOREM PROVER,[91] *before* a new function symbol f_k is admitted to the specification, a "valid induction template" — which immediately implies termination — has to be constructed from the defining equation of f_k.[92]

Induction templates were first used in THM and received their name when they were first described in [Boyer & Moore, 1979].

Every time a new recursive function f_k is defined, a system for explicit induction immediately tries to construct *valid induction templates*; if it does not find any, then the new function symbol is rejected w.r.t. the given definition; otherwise the system links the function name with its definition and its valid induction templates.

The induction templates serve actually two purposes: as witnesses for termination and as the basic tools of the induction rule of explicit induction for generating the step cases.

For a finite number of mutually recursive functions f_k with arity n_k ($k \in K$), an induction template in the most general form consists of the following:

1. A *relational description*[93] of the changes in the argument pattern of these recursive functions as found in their recursive defining equations:
   For each $k \in K$ and for each positive/negative-conditional equation with a left-hand side of the form $f_k(t_1, \ldots, t_{n_k})$, we take the set R of all recursive function calls of the $f_{k'}$ ($k' \in K$) occurring in the right-hand side or the condition, and some case condition C, which must be a subset of the conjunctive condition literals of the defining equation. Typically, C is empty (i.e. always true) in the case of constructor-style definitions, and just sufficient to guarantee proper destructor applications in the case of destructor-style definitions.
   Together they form the triple $(f_k(t_1, \ldots, t_{n_k}), R, C)$, and a set containing such a triple for each such defining equation forms the relational description.

---

[91] Note that termination is not proved in the PURE LISP THEOREM PROVER; instead, the soundness of the induction proofs comes with the *proviso* that the rewrite relation of all defined function symbols terminate.

[92] See also item (d) of the "definition principle" of [Boyer & Moore, 1979, p. 44f.] for a formulation that avoids the technical term "induction template".

[93] The name "relational description" comes from [Walther, 1992; 1993].

For our definition of + via (+1), (+2) in §4.4, there is only one recursive equation and only one relevant relational description, namely the following one with an empty case condition:
$$\{\ (\ \mathsf{s}(x)+y,\ \{x+y\},\ \emptyset\ )\ \}.$$
Also for our definition of + with (+1′), (+2′) in Example 5.1, there is only one recursive equation and only one relevant relational description, namely
$$\{\ (\ x+y,\ \{\mathsf{p}(x)+y\},\ \{x\neq 0\}\ )\ \}.$$

2. For each $k \in K$, a variable-free weight term w_{f_k} in which the position numbers
$$(1),\ldots,(n_k)$$
are used in place of variables. The position numbers actually occurring in the term are called the *measured positions*.

For our two relational descriptions, only the weight term (1) (consisting just of a position number) makes sense as w_+, resulting in the set of measured positions $\{1\}$. Indeed, + terminates in both definitions because the argument in the first position gets smaller.

3. A binary predicate < that is known to represent a well-founded relation.

For our two relational descriptions, the predicate $\lambda x,y.\ (\mathsf{lessp}(x,y) = \mathsf{true})$ is appropriate.

Now, an induction template is *valid* if for each element of the relational description as given above, and for each $f_{k'}(t'_1,\ldots,t'_{n_{k'}}) \in R$, the following conjecture is valid:
$$w_{f_{k'}}\{(1) \mapsto t'_1, \ldots, (n_{k'}) \mapsto t'_{n_{k'}}\}\ <\ w_{f_k}\{(1) \mapsto t_1, \ldots, (n_k) \mapsto t_{n_k}\}\ \Leftarrow\ \bigwedge C.$$
For our two relational descriptions, this amounts to showing $\mathsf{lessp}(x,\mathsf{s}(x)) = \mathsf{true}$ and $\mathsf{lessp}(\mathsf{p}(x),x) = \mathsf{true} \Leftarrow x \neq 0$, respectively; so their templates are both valid by lemma (lessp4) and axioms (nat1–2) and (p1).

Example 5.2 (Two Induction Templates, Different Measured Positions)
For the ordering predicate lessp as defined by (lessp1–3) of §4.4, we get two appropriate induction templates with the sets of measured positions $\{1\}$ and $\{2\}$, respectively, both with the relational description
$$\{\ (\ \mathsf{lessp}(\mathsf{s}(x),\mathsf{s}(y)),\ \{\mathsf{lessp}(x,y)\},\ \emptyset\ )\ \},$$
and both with the well-founded ordering $\lambda x,y.\ (\mathsf{lessp}(x,y) = \mathsf{true})$. The first template has the weight term (1) and the second one has the weight term (2). The validity of both templates is given by lemma (lessp4) of §4.4. □

Example 5.3 (One Induction Template with Two Measured Positions)
For the Ackermann function ack as defined by (ack1–3) of §4.4, we get only one appropriate induction template. The set of its measured positions is $\{1,2\}$, because of the weight function $\mathsf{cons}((1), \mathsf{cons}((2), \mathsf{nil}))$, which we will abbreviate in the following with $[(1), (2)]$. The well-founded relation is the lexicographic ordering $\lambda l, k.$ $(\mathsf{lexlimless}(l, k, \mathsf{s}(\mathsf{s}(\mathsf{s}(0)))) = \mathsf{true})$. The relational description has two elements: For the equation (ack2) we get
$$(\ \mathsf{ack}(\mathsf{s}(x), 0), \ \ \{\mathsf{ack}(x, \mathsf{s}(0))\}, \ \ \emptyset \ ),$$
and for the equation (ack3) we get
$$(\ \mathsf{ack}(\mathsf{s}(x), \mathsf{s}(y)), \ \ \{\mathsf{ack}(\mathsf{s}(x), y), \ \mathsf{ack}(x, \mathsf{ack}(\mathsf{s}(x), y))\}, \ \ \emptyset \ ).$$
The validity of the template is expressed in the three equations

$\mathsf{lexlimless}([x, \mathsf{s}(0)],$	$[\mathsf{s}(x), 0],$	$\mathsf{s}(\mathsf{s}(\mathsf{s}(0))))$	$=$	true;
$\mathsf{lexlimless}([\mathsf{s}(x), y],$	$[\mathsf{s}(x), \mathsf{s}(y)], \mathsf{s}(\mathsf{s}(\mathsf{s}(0))))$		$=$	true;
$\mathsf{lexlimless}([x, \mathsf{ack}(\mathsf{s}(x), y)], [\mathsf{s}(x), \mathsf{s}(y)], \mathsf{s}(\mathsf{s}(\mathsf{s}(0))))$			$=$	true;

which follow deductively from (lessp4), (lexlimless1), (lexless2–4), (length1–2). □

For induction templates of destructor-style definitions see Examples 6.8 and 6.9 in §6.3.7.

5.6 Termination of the Rewrite Relation on Ground Terms

Let us prove that the existence of a valid induction template for a new set of recursive functions f_k ($k \in K$) actually implies termination of the rewrite relation after addition of the new positive/negative-conditional equations for the f_k, assuming an arbitrary free-constructor model $\mathcal{M}$ of all (old and new) (positive/negative-conditional) equations to be given.[94]

For an *argumentum ad absurdum*, suppose that there is an infinite sequence of rewrite steps on ground terms. Consider each term in this sequence to be replaced with the multiset that contains, for each occurrence of a function call $f_k(t_1, \ldots, t_{n_k})$ with $k \in K$, the value of its weight term $w_{f_k}\{(1) \mapsto t_1, \ldots, (n_k) \mapsto t_{n_k}\}$ in $\mathcal{M}$.

Then the rewrite steps with instances of the *old* equations of previous function definitions (of symbols not among the f_k) can change the multiset only by deleting some elements for the following two reasons: Instances that do not contain any new function symbol have no effect on the values in $\mathcal{M}$, because $\mathcal{M}$ is a model of the old equations. There are no other instances because the new function symbols

---
[94] A *free-constructor model* is a model where two constructor ground terms are equal in $\mathcal{M}$ only if they are syntactically equal. Because the confluence result of [Wirth, 2009] applies in our case without requiring termination, there is always an initial free-constructor model according to Corollary 7.17 of [Wirth, 1997], namely the factor algebra of the ground term algebra modulo the equivalence closure of the rewrite relation.

do not occur in the old equations, and because we consider all our variables to be constructor variables as explained in § 5.4.[95]

Moreover, a rewrite step with a *new* equation reduces only a single innermost occurrence of a new function symbol, because only a single new function symbol occurs on the left-hand side of the equation and because we consider all our variables to be constructor variables. The other occurrences in the multiset are not affected because $\mathcal{M}$ is a model of the new equations. Thus, such a rewrite step reduces the multiset in a well-founded relation, namely the multiset extension of the well-founded relation of the template in the assumed model $\mathcal{M}$. Indeed, this follows from the fulfilledness of the conditions of the equation and the validity of the template.

Thus, in each rewrite step, the multiset gets smaller in a well-founded ordering or does not change. Moreover, if we assume that rewriting with the old equations terminates, then the new equations must be applied infinitely often in this sequence, and so the multiset gets smaller in infinitely many steps, which is impossible in a well-founded ordering.

5.7 Applicable Induction Templates for Explicit Induction

We restrict the discussion in this section to recursive functions that are not mutually recursive, partly for simplicity and partly because induction templates are hardly helpful for finding proofs involving non-trivially mutually recursive functions.[96]

Moreover, in principle, users can always encode mutually recursive functions $f_k(\ldots)$ by means of a single recursive function $f(k, \ldots)$. Via such an encoding, humans tend to provide additional heuristic information relevant for induction templates, namely by the way they standardize the argument list w.r.t. length and position (cf. the "changeable positions" below).

Thus, all the f_k with arity n_k of § 5.5 simplify to one symbol f with arity n. Under this restriction it is easy to partition the measured positions of a template into "changeable" and "unchangeable" ones.[97]

---

[95] Among the old equations here, we may even admit projective equations with *general* variables, such as for destructors and the conditional function IfThenElse$_{nat}$: bool, nat, nat → nat:

$p(s(X)) = X$ | $car(cons(X, L)) = X$ | IfThenElse$_{nat}$(true, X, Y) = X
 | $cdr(cons(X, L)) = L$ | IfThenElse$_{nat}$(false, X, Y) = Y

for general variables X, Y : nat, L : list(nat), ranging over general terms (instead of constructor terms only).

[96] See, however, [Kapur & Subramaniam, 1996] for explicit-induction heuristics applicable to simple forms of mutual recursion.

[97] This partition into changeable and unchangeable positions (actually: variables) originates in [Boyer & Moore, 1979, p. 185f.].

Changeable are those measured positions i of the template which sometimes change in the recursion, i.e. for which there is a triple $(f(t_1,\ldots,t_n), R, C)$ in the relational description of the template, and an $f(t'_1,\ldots,t'_n) \in R$ such that $t'_i \neq t_i$. The remaining measured positions of the template are called *unchangeable*. Unchangeable positions typically result from the inclusion of a global variable into the argument list of a function (to observe an applicative programming style).

To improve the applicability of the induction hypotheses of the step cases produced by the induction rule, these induction hypotheses should mirror the recursive calls of the unfolding of the definition of a function f occurring in the induction rule's input formula, say
$$A[f(t''_1,\ldots,t''_n)].$$

An induction template is *applicable* to the indicated occurrence of its function symbol f if the terms t''_i at the changeable positions i of the template are *distinct variables* and none of these variables occurs in the terms $t''_{i'}$ that fill the unchangeable positions i' of the template.[98] For templates of constructor-style equations we additionally have to require here that the first element $f(t_1,\ldots,t_n)$ of each triple of the relational description of the template matches $(f(t''_1,\ldots,t''_n))\xi$ for some *constructor substitution* ξ that may replace the variables of $f(t''_1,\ldots,t''_n)$ with constructor terms, i.e. terms consisting of constructor symbols and variables, such that $t''_i\xi = t''_i$ for each unchangeable position i of the template.

Example 5.4 (Applicable Induction Templates)

Let us consider the conjecture (ack4) from §4.4. From the three induction templates of Examples 5.2 and 5.3, only the one of Example 5.3 is applicable. The two of Example 5.2 are not applicable because $\mathsf{lessp}(\mathsf{s}(x), \mathsf{s}(y))$ cannot be matched to $(\mathsf{lessp}(y, \mathsf{ack}(x,y)))\xi$ for any constructor substitution ξ. □

5.8 Induction Schemes

Let us recall that for every recursive call $f(t'_{j',1},\ldots,t'_{j',n})$ in a positive/negative-conditional equation with left-hand side $f(t_1,\ldots,t_n)$, the relational description of an induction template for f contains a triple
$$\bigl(f(t_1,\ldots,t_n),\ \{ f(t'_{j,1},\ldots,t'_{j,n}) \mid j \in J \},\ C \bigr),$$
such that $j' \in J$ (by definition of an induction template).

Let us assume that the induction template is valid and applicable to the occurrence indicated in the formula $A[f(t''_1,\ldots,t''_n)]$ given as input to the induction rule of explicit induction. Let σ be the substitution whose domain are the variables

---

[98] This definition of applicability originates in [Boyer & Moore, 1979, p. 185f.].

of $f(t_1,\ldots,t_n)$ and which matches the first element $f(t_1,\ldots,t_n)$ of the triple to $(f(t_1'',\ldots,t_n''))\xi$ for some constructor substitution ξ whose domain are the variables of $f(t_1'',\ldots,t_n'')$, such that $t_i''\xi = t_i''$ for each unchangeable position i of the template. Then we have $t_i\sigma = t_i''\xi$ for $i \in \{1,\ldots,n\}$.

Now, for the well-foundedness of the generic step-case formula

$$\Big(\big(A[f(t_1'',\ldots,t_n'')]\big)\xi \;\;\Leftarrow\;\; \bigwedge\nolimits_{j\in J}\big(A[f(t_1'',\ldots,t_n'')]\big)\mu_j \Big) \;\;\Leftarrow\;\; \bigwedge C\sigma$$

to be implied by the validity of the induction template, it suffices (because of $t_i''\xi = t_i\sigma$) to take substitutions μ_j whose domain $\mathrm{dom}(\mu_j)$ is the set of variables of $f(t_1'',\ldots,t_n'')$, such that the *matching constraint*

$$t_i''\mu_j = t_{j,i}'\sigma$$

is satisfied for each measured position i of the template and for each $j \in J$.

If i is an unchangeable position of the template, then we have $t_i = t_{j,i}'$ and $t_i''\xi = t_i''$. Therefore, we can satisfy the matching constraint by requiring μ_j to be the identity on the variables of t_i'', simply because then we have $t_i''\mu_j = t_i'' = t_i''\xi = t_i\sigma = t_{j,i}'\sigma$.

If i is a changeable position, then we know by the applicability of the template that t_i'' is a variable not occurring in another changeable or unchangeable position in $f(t_1'',\ldots,t_n'')$, and we can satisfy the matching constraint simply by defining $t_i''\mu_j := t_{j,i}'\sigma$.

On the remaining variables of $f(t_1'',\ldots,t_n'')$, we define μ_j in a way that we get $t_i''\mu_j = t_{j,i}'\sigma$ for as many unmeasured positions i as possible, and otherwise as the identity. This is not required for well-foundedness, but it improves the likeliness of applicability of the induction hypothesis $(A[f(t_1'',\ldots,t_n'')])\mu_j$ after unfolding $f(t_1'',\ldots,t_n'')\xi$ in $(A[f(t_1'',\ldots,t_n'')])\xi$.

Note that such an eager instantiation of the input formula via $\{\mu_j \mid j \in J\}$ is required in explicit induction unless the logic admits one of the following: existential quantification, existential variables,[99] lazy induction-hypothesis generation.

An *induction scheme* for the given input formula consists of the following items:

1. The *position set* contains the position of $f(t_1'',\ldots,t_n'')$ in $A[f(t_1'',\ldots,t_n'')]$. Merging of induction schemes may lead to non-singleton position sets later.

2. The set of the *induction variables*, which are defined as the variables at the changeable positions of the induction template in $f(t_1'',\ldots,t_n'')$.

3. To obtain a *step-case description* for all step cases by means of the generic step-case formula displayed above, each triple in the relational description of the considered form is replaced with the new triple
$$\big(\; \xi,\; \{\mu_j \mid j \in J\},\; C\sigma \;\big).$$

To make as many induction hypotheses available as possible in each case, we assume that step-case descriptions are implicitly kept normalized by the following associative commutative operation: If two triples are identical in their first elements and in their last elements, we replace them with the single triple that has the same first and last elements and the union of the middle elements as new middle element.

4. We also add the *hitting ratio* of all substitutions μ_j with $j \in J$ given by
$$\frac{|\{\,(j,i) \in J \times \{1,\ldots,n\} \mid t''_i \mu_j = t'_{j,i}\sigma\,\}|}{|J \times \{1,\ldots,n\}|},$$
where J actually has to be the disjoint sum over all the J occurring as index sets of second elements of triples like the one displayed above. We newly introduce the name "hitting ratio" here in the hope that it helps the readers to remember that this ratio measures how well the induction hypotheses hit the recursive calls according the matching constraint displayed before.

Note that the resulting step-case description is a set describing all step cases of an induction scheme; these step cases are guaranteed to be well-founded,[100] but — for providing a sound induction formula — they still have to be complemented by base cases, which may be analogously described by triples $(\xi, \emptyset, C)$, such that all substitutions in the first elements of the triples together describe a distinction of cases that is complete for constructor terms and, for each of these substitutions, its case conditions describe a complete distinction of cases again.

Example 5.5 (Induction Scheme)

The template for ack of Example 5.3 is the only one that is applicable to (ack4) according to Example 5.4. It yields the following induction scheme.

The *position set* is $\{1.1.2\}$. It describes the occurrence of ack in the second subterm of the left-hand side of the first literal of the formula (ack4) as input to the induction rule of explicit induction:

$$(\text{ack4}) \,/\, 1.1.2 \;=\; \text{ack}(x,y).$$

The set of *induction variables* is $\{x, y\}$, because both positions of the induction template are changeable.

---

[99]Existential variables are called "free variables" in modern tableau systems (cf. [Fitting, 1990; 1996]) and occur with extended functionality under different names in the inference systems of [Wirth, 2004; 2012b; 2017].

[100]Well-foundedness is indeed guaranteed according to the above discussion. As a consequence, the induction scheme does not need the weight term and the well-founded relation of the induction template anymore.

The relational description of the induction template is replaced with the *step-case description*
$$\{\ (\ \xi_1,\ \{\mu_{1,1}\},\ \emptyset\ ),\quad (\ \xi_2,\ \{\mu_{2,1}, \mu_{2,2}\},\ \emptyset\ )\ \}.$$
that is given as follows.

The first triple of the relational description, namely
$$(\ \mathsf{ack}(\mathsf{s}(x), 0),\ \{\mathsf{ack}(x, \mathsf{s}(0))\},\ \emptyset\ )$$
(obtained from the equation (ack2)) is replaced with
$$(\ \xi_1,\ \{\mu_{1,1}\},\ \emptyset\ ),$$
where
$$\xi_1 = \{x \mapsto \mathsf{s}(x'),\ y \mapsto 0\}\ \text{and}$$
$$\mu_{1,1} = \{x \mapsto x',\ y \mapsto \mathsf{s}(0)\}.$$
This can be seen as follows. The substitution called σ in the above discussion — which has to match the first element of the triple to $((\mathsf{ack4})/1.1.2)\xi_1$ — has to satisfy $(\mathsf{ack}(\mathsf{s}(x), 0))\sigma = (\mathsf{ack}(x, y))\xi_1$. Taking ξ_1 as the minimal constructor substitution given above, this determines $\sigma = \{x \mapsto x'\}$. Moreover, as both positions of the template are changeable, $\mu_{1,1}$ has to match $(\mathsf{ack4})/1.1.2$ to the σ-instance of the single element of the second element of the triple, which determines $\mu_{1,1}$ as given.

The second triple of the relational description, namely
$$(\ \mathsf{ack}(\mathsf{s}(x), \mathsf{s}(y)),\ \{\mathsf{ack}(\mathsf{s}(x), y),\ \mathsf{ack}(x, \mathsf{ack}(\mathsf{s}(x), y))\},\ \emptyset\ )$$
(obtained from the equation (ack3)) is replaced with $(\ \xi_2,\ \{\mu_{2,1}, \mu_{2,2}\},\ \emptyset\ )$, where
$$\xi_2 = \{x \mapsto \mathsf{s}(x'),\ y \mapsto \mathsf{s}(y')\},$$
$$\mu_{2,1} = \{x \mapsto \mathsf{s}(x'),\ y \mapsto y'\},\ \text{and}$$
$$\mu_{2,2} = \{x \mapsto x',\ y \mapsto \mathsf{ack}(\mathsf{s}(x'), y')\}.$$
This can be seen as follows. The substitution called σ in the above discussion has to satisfy $(\mathsf{ack}(\mathsf{s}(x), \mathsf{s}(y)))\sigma = (\mathsf{ack}(x, y))\xi_2$. Taking ξ_2 as the minimal constructor substitution given above, this determines $\sigma = \{x \mapsto x',\ y \mapsto y'\}$. Moreover, we get the constraints $(\mathsf{ack}(x, y))\mu_{2,1} = (\mathsf{ack}(\mathsf{s}(x), y))\sigma$ and $(\mathsf{ack}(x, y))\mu_{2,2} = (\mathsf{ack}(x, \mathsf{ack}(\mathsf{s}(x), y)))\sigma$, which determine $\mu_{2,1}$ and $\mu_{2,2}$ as given above.

The hitting ratio for the three constraints on the two arguments of $(\mathsf{ack4})/1.1.2$ is $\frac{6}{6} = 1$. This is optimal: the induction hypotheses are 100% identical to the expected recursive calls.

To achieve completeness of the substitutions ξ_k for constructor terms we have to add the base case $(\xi_0, \emptyset, \emptyset)$ with $\xi_0 = \{x \mapsto 0,\ y \mapsto y\}$ to the step-case description.

The three new triples now describe exactly the three formulas displayed at the beginning of Example 4.5 in § 4.9. □

6 Automated Explicit Induction

6.1 The Application Context of Automated Explicit Induction

Since the development of programmable computing machinery in the middle of the 20th century, a major problem of hard- and software has been and still is the uncertainty that they actually always do what they should do.

It is almost never the case that the product of the possible initial states, input threads, and schedulings of a computing system is a small number. Otherwise, however, even the most carefully chosen test series cannot cover the often very huge or even infinite number of possible cases; and then, no matter how many bugs have been found by testing, there can never be certainty that none remain.

Therefore, the only viable solution to this problem seems to be:

> Specify the intended functionality in a language of formal logic, and then supply a formal mechanically checked proof that the program actually satisfies the specification!

Such an approach also requires formalizing the platforms on which the system is implemented. This may include the hardware, operating system, programming language, sensory input, etc. One may additionally formalize and prove that the underlying platforms are implemented correctly and this may ultimately involve proving, for example, that a network of logical gates and wires implements a given abstract machine. Eventually, however, one must make an engineering judgment that certain physical objects (e.g. printed circuit boards, gold plated pins, power supplies, etc.) reliably behave as specified. To be complete, such an approach would also require a verification that the verification system is sound and correctly implemented.[101]

A crucial problem, however, is the cost — in time and money — of doing the many proofs required, given the huge amounts of application hard- and software in our modern economies. Thus, we can expect formal verification only in areas where the managers expect that mere testing does not suffice, that the costs of the verification process are lower than the costs of bugs in the hard- or software, and that the competitive situation admits the verification investment. Good candidates are the areas of central processing units (CPUs) in standard processors and of security protocols.

---

[101] See, for example, [Davis, 2009].

To reduce the costs of verification, we can hope to automate it with automated theorem-proving systems. This automation has to include mathematical induction because induction is essential for the verification of the properties of most data types used in digital design (such as natural numbers, arrays, lists, and trees), for the repetition in processing (such as loops), and for parameterized systems (such as a generic n-bit adder).

Decision methods (many of them exploiting finiteness, e.g. the use of 32-bit data paths) allow automatic verification of some modules, but — barring a completely unexpected breakthrough in the future — the verification of a new hard- or software system will always require human users who help the theorem-proving systems to explore and develop the notions and theories that properly match the new system.

Already today, however, ACL2 often achieves complete automation in verifying minor modifications of previously verified modules — an activity called *proof maintenance* which is increasingly important in the microprocessor-design industry.

6.2 The PURE LISP THEOREM PROVER

Our overall task is to answer — from a historical perspective — the question:

> How did Robert S. Boyer and J Strother Moore — starting virtually from zero[102] in the summer of 1972 — invent their long-lived solutions to the hard heuristic problems in the automation of induction and implement them in the sophisticated theorem prover THM as described in [Boyer & Moore, 1979]?

---

[102] No heuristics at all were explicitly described, for instance, in Burstall's 1968 work on program verification by induction over recursive functions in [Burstall, 1969], where the proofs were not even formal, and an implementation seemed to be more or less utopian [Burstall, 1969, p. 41]:

"The proofs presented will be mathematically rigorous but not formalised to the point where each inference is presented as a mechanical application of elementary rules of symbol manipulation. This is deliberate since I feel that our first aim should be to devise methods of proof which will prove the validity of non-trivial programs in a natural and intelligible manner. Obviously we will wish at some stage to formalise the reasoning to a point where it can be performed by a computer to give a mechanised debugging service."

As far as we are aware, besides interactively invoked induction in resolution theorem proving (e.g. by starting a resolution proof for the two clauses resulting from Skolemization of $(P(0) \land \neg P(x)) \Rightarrow \exists y.\ (P(y) \land \neg P(s(y)))$ [Darlington, 1968]), the only implementation of an automatically invoked mathematical-induction heuristic prior to 1972 is in a set-theory prover by Bledsoe [1971], which uses structural induction over 0 and s (cf. § 4.4) on a randomly picked, universally quantified variable of type nat.

As already described in §1, the breakthrough in the heuristics for automated inductive theorem proving was achieved with the "PURE LISP THEOREM PROVER", developed and implemented by Boyer and Moore. It was presented by Moore at the third IJCAI [Boyer & Moore, 1973], which took place in Stanford (CA) in August 1973, and it is best documented in Part II of Moore's PhD thesis [1973], defended in November 1973.

The PURE LISP THEOREM PROVER was given no name in the before-mentioned publications. The only occurrence of the name in publication seems to be in [Moore, 1975a, p. 1], where it is actually called "the Boyer–Moore PURE LISP THEOREM PROVER".

To make a long story short, the fundamental insights were

- to exploit the duality of recursion and induction to formulate explicit induction hypotheses,
- to abandon "random" search and focus on simplifying the goal by rewriting and normalization techniques to lead to opportunities to use the induction hypotheses, and
- to support generalization to prepare subgoals for subsequent inductions.

Thus, it is not enough for us to focus here just on the induction heuristics *per se*, but it is necessary to place them in the context of the development of the Boyer–Moore waterfall (cf. Figure 1).

To understand the achievements a bit better, let us now discuss the material of Part II of Moore's PhD thesis in some detail, because it provides some explanation of how Boyer and Moore could be so surprisingly successful. Especially helpful for understanding the process of creation are those procedures of the PURE LISP THEOREM PROVER that are provisional w.r.t. their refinement in later Boyer–Moore theorem provers. Indeed, these provisional procedures help to decompose the leap from nothing to THM, which was achieved by two men in less than eight years of work.

As W. W. Bledsoe (1921–1995) was Boyer's PhD advisor, it is no surprise that the PURE LISP THEOREM PROVER shares many design features with Bledsoe's provers. In [Moore, 1973, p.172], we read on the PURE LISP THEOREM PROVER:

> "The design of the program, especially the straightforward approach of 'hitting' the theorem over and over again with rewrite rules until it can no longer be changed, is largely due to the influence of W. W. Bledsoe."

Boyer and Moore report[103] that in late 1972 and early 1973 they were doing proofs about list data structures on the blackboard and verbalizing to each other the heuristics behind their choices on how to proceed with the proof. This means that, although explicit induction is not the approach humans would choose for nontrivial induction tasks, the heuristics of the PURE LISP THEOREM PROVER are learned from human heuristics after all.

Note that Boyer and Moore's method of learning computer heuristics from their own human behavior in mathematical logic was a step of two young men against the spirit of the time: the use of vast amounts of computational power to *search* an even more enormous space of possibilities. Boyer and Moore's goal, however, was in a sense more modest:

> "The program was designed to behave properly on simple functions. The overriding consideration was that it should be automatically able to prove theorems about simple LISP functions in the straightforward way we prove them." [Moore, 1973, p. 205]

It may be that the orientation toward human-like or "intelligible" methods and heuristics in the automation of theorem proving had also some tradition in Edinburgh at the time,[104] but, also in this aspect, the major influence on Boyer and Moore is again W. W. Bledsoe.[105]

The source code of the PURE LISP THEOREM PROVER was written in the programming language POP–2.[106] Boyer and Moore were the only programmers involved in the implementation. The average time in the central processing unit (CPU) of the ICL–4130 for the proof of a theorem is reported to be about ten seconds.[107] This was considered fast at the time, compared to the search-dominated proofs by resolution systems. Moore explains the speed:

> "Finally, it should be pointed out that the program uses no search. At no time does it 'undo' a decision or back up. This is both the primary reason it is a fast theorem prover, and strong evidence that its methods allow the theorem to be proved in the way a programmer might 'observe' it. The program is designed to make the right guess the first time, and then pursue one goal with power and perseverance." [Moore, 1973, p. 208]

---

[103] Cf. [Wirth, 2012d].

[104] Cf. e.g. the quotation from [Burstall, 1969] in Note 102.

[105] Cf. e.g. [Bledsoe &al., 1972].

[106] Cf. [Burstall &al., 1971].

One remarkable omission in the PURE LISP THEOREM PROVER is lemma application. As a consequence, the success of proving a set of theorems cannot depend on the order of their presentation to the theorem prover. Indeed, just as the resolution theorem provers of the time, the PURE LISP THEOREM PROVER starts every proof right from scratch and does not improve its behavior with the help of previously proved lemmas. This was a design decision; one of the reasons was:

> "Finally, one of the primary aims of this project has been to demonstrate clearly that it is possible to prove program properties entirely automatically. A total ban on all built-in information about user defined functions thus removes any taint of user supplied information."
>
> [Moore, 1973, p. 203]

Moreover, all induction orderings in the PURE LISP THEOREM PROVER are recombinations of constructor relations, such that all inductions it can do are structural inductions over combinations of constructors. As a consequence, contrary to later Boyer–Moore theorem provers, the well-foundedness of the induction orderings does not depend on the termination of the recursive function definitions.[108]

Nevertheless, the soundness of the PURE LISP THEOREM PROVER depends on the termination of the recursive function definitions, but only in one aspect: It simplifies and evaluates expressions under the assumption of termination. For instance, both (IF[109] a d d) and (CDR (CONS a d)) simplify to d, no matter whether a terminates; and it is admitted to rewrite with a recursive function definition even if an argument of the function call does not terminate. Note that such a lazy form of evaluation is sound w.r.t. the given logic only if each eager call terminates and returns a constructor ground term, simply because all functions are meant to be defined in terms of constructor variables (cf. § 5.4).[110]

The termination of the recursively defined functions, however, is not checked by the PURE LISP THEOREM PROVER, but comes as a *proviso* for its soundness.

---

[107] Here is the actual wording of the timing result found on Page 171f. of [Moore, 1973]: "Despite theses inefficiencies, the 'typical' theorem proved requires only 8 to 10 seconds of CPU time. For comparison purposes, it should be noted that the time for CONS in 4130 POP–2 is 400 microseconds, and CAR and CDR are about 50 microseconds each. The hardest theorems solved, such as those involving SORT, require 40 to 50 seconds each."

[108] Note that the well-foundedness of the constructor relations depends on distinctness of the constructor ground terms in the models, but this does not really depend on the termination of the recursive functions because (as discussed in § 5.2) confluence is sufficient here.

[109] Cf. Note 90.

[110] There is a work-around for projective functions as indicated in Note 95 and in [Wirth, 2009].

The logic of the PURE LISP THEOREM PROVER is an applicative[111] subset of the logic of LISP. The only *destructors* in this logic are CAR and CDR. They are overspecified on the only *constructors* NIL and CONS by the following equations:

$$\begin{array}{rcl|rcl} (\text{CAR (CONS } a \ d)) & = & a & (\text{CAR NIL}) & = & \text{NIL} \\ (\text{CDR (CONS } a \ d)) & = & d & (\text{CDR NIL}) & = & \text{NIL} \end{array}$$

As standard in LISP, every term of the form (CONS a d) is taken to be **true** in the logic of the PURE LISP THEOREM PROVER if it occurs at an argument position with Boolean intention. The actual truth values (to be returned by Boolean functions) are NIL (representing **false**) and T, which is an abbreviation for (CONS NIL NIL) and represents **true**.[112] Unlike conventional LISPs (both then and now), the natural numbers are represented by lists of NILs to keep the logic simple; the natural number 0 is represented by NIL and the successor function s(d) is represented by (CONS NIL d).[113]

Let us now discuss the behavior of the PURE LISP THEOREM PROVER by describing the instances of the stages of the Boyer–Moore waterfall (cf. Figure 1) as they are described in Moore's PhD thesis.

6.2.1 Simplification in the PURE LISP THEOREM PROVER

The first stage of the Boyer–Moore waterfall — "simplification" in Figure 1 — is called "normalation"[114] in the PURE LISP THEOREM PROVER. It applies the following simplification procedures to LISP expressions until the result does not change any more: "*evaluation*", "*normalization*", and "*reduction*".

"*Evaluation*" is a procedure that evaluates expressions partly by simplification within the elementary logic as given by Boolean operations and the equality predicate. Moreover, "evaluation" executes some rewrite steps with the equations defining the recursive functions. Thus, "evaluation" can roughly be seen as normalization with

---

[111] Cf. Note 82.

[112] Cf. 2nd paragraph of Page 86 of [Moore, 1973].

[113] Cf. 2nd paragraph of Page 87 of [Moore, 1973].

[114] During the oral defense of the dissertation, Moore's committee abhorred the non-word and instructed him to choose a word. Some copies of the dissertation call the process "simplification."

the rewrite relation resulting from the elementary logic and from the recursive function definitions. The rewrite relation is applied according to the innermost left-to-right rewriting strategy, which is standard in LISP.

"Evaluation" completely evaluates all ground terms to their normal forms. Terms containing (implicitly universally quantified) variables, however, have to be handled in addition. Surprisingly, the considered rewrite relation is not necessarily terminating on non-ground terms, although the LISP evaluation of ground terms terminates because of the assumed termination of recursive function definitions (cf. §5.5). The reason for this non-termination is the following: Because of the LISP definition style via *un*conditional equations, the positive/negative conditions are actually part of the *right-hand sides* of the defining equations, such that the rewrite step can be executed even if the conditions evaluate neither to false nor to true. For instance, in Example 5.1 of §5.2, a rewrite step with the definition of PLUS can always be executed, whereas a rewrite step with $(+1')$ or $(+2')$ requires $x=0$ to be definitely true or definitely false. This means that non-termination may result from the rewriting of cases that do not occur in the evaluation of any ground instance.[115]

As the final aim of the stages of the Boyer–Moore waterfall is a formula that provides concise and sufficiently strong induction hypotheses in the last of these stages, symbolic evaluation must be prevented from unfolding function definitions unless the context admits us to expect an effect of simplification.[116]

Because the main function of "evaluation" — only to be found in this first one of the Boyer–Moore theorem provers — is to collect data to assist the induction rule in the generation of appropriate base and step cases later, the PURE LISP THEOREM PROVER applies a unique procedure to stop the unfolding of recursive function definitions:

---

[115]It becomes clear in the second paragraph on Page 118 of [Moore, 1973] that the code of both the positive and the negative case of a conditional will be evaluated, unless one of them can be canceled by the complete evaluation of the governing condition to true or false. Note that the evaluation of both cases is necessary indeed and cannot be avoided in practice.

Moreover, note that a stronger termination requirement that guarantees termination independent of the governing condition is not feasible for recursive function definitions in practice.

Later Boyer–Moore theorem provers also use lemmas for rewriting during symbolic evaluation, which is another source of possible non-termination.

The mechanism for partially enforcing termination of "evaluation" according to this procedure is vaguely described in the last paragraph on Page 118 of Moore's PhD thesis. As this kind of "evaluation" is only an intermediate solution on the way to more refined control information for the induction rule in later Boyer–Moore theorem provers, the rough information given here may suffice.

[116]In QUODLIBET this is achieved by *contextual rewriting* where evaluation stops when the governing conditions cannot be established from the context. Cf. [Schmidt-Samoa, 2006b; 2006c].

A rewrite step with an equation defining a recursive function f is canceled if there is a CAR or a CDR in an argument to an occurrence of f in the right-hand side of the defining equation that is encountered during the control flow of "evaluation", and if this CAR or CDR is not removed by the "evaluation" of the arguments of this occurrence of f under the current environment updated by matching the left-hand side of the equation to the redex. For instance, "evaluation" of (PLUS (CONS NIL X) Y) returns (CONS NIL (PLUS X Y)); whereas "evaluation" of (PLUS X Y) returns (PLUS X Y) and informs the induction rule that (CDR X) occurred in the recursive call during the trial to rewrite with the definition of PLUS. In general, such occurrences indicate which induction hypotheses should be generated by the induction rule.[117]

"Evaluation" provides a crucial link between symbolic evaluation and the induction rule of explicit induction. The question "Which case distinction on which induction variables should be used for the induction proof and how should the step cases look?" is reduced to the quite different question "Where do destructors like CAR and CDR heap up during symbolic evaluation?". This reduction helps to understand by which intermediate steps it was possible to develop the most surprising, sophisticated recursion analysis of later Boyer–Moore theorem provers.

"*Normalization*" tries to find sufficient conditions for a given expression to have the soft type "Boolean" and to normalize logical expressions. Contrary to clausal logic over equational atoms, LISP admits EQUAL and IF to appear not only at the top level, but in nested terms. To free later tests and heuristics from checking for their triggers in every equivalent form, such a normalization w.r.t. propositional logic and equality is part of most theorem provers today.

"*Reduction*" is a rudimentary form of what today is called *contextual rewriting*. It is based on the fact that — in the logic of the PURE LISP THEOREM PROVER — in the conditional expression

(IF c p n)

we can simplify occurrences of c in p to (CONS (CAR c) (CDR c)), and in n to NIL. The replacement with (CONS (CAR c) (CDR c)) is executed only at positions with Boolean intention and can be improved in the following two special cases:

---

[117] Actually, "evaluation" also informs which occurrences of CAR or CDR besides the arguments of recursive occurrences of PLUS were permanently introduced during that trial to rewrite. Such occurrences trigger an additional case analysis to be generated by the induction rule, mostly as a compensation for the omission of the stage of "destructor elimination" in the PURE LISP THEOREM PROVER.

1. If we know that c is of soft type "Boolean", then we rewrite all occurrences of c in p actually to T.

2. If c is of the form (EQUAL l r), then we can rewrite occurrences of l in p to r (or vice versa). Note that we have to treat the variables in l and r as constants in this rewriting. The PURE LISP THEOREM PROVER rewrites in this case only if either l or r is a ground term;[118] then the other cannot be a ground term because the equation would otherwise have been simplified to T or NIL in the previously applied "evaluation". So replacing the latter term with the ground term everywhere in p must terminate, and this is all the contextual rewriting with equalities that the PURE LISP THEOREM PROVER does in "reduction".[119]

6.2.2 Destructor Elimination in the PURE LISP THEOREM PROVER

There is no such stage in the PURE LISP THEOREM PROVER.[120]

6.2.3 (Cross-) Fertilization in the PURE LISP THEOREM PROVER

Fertilization is just contextual rewriting with an equality, as described before (cf. the "reduction" that is part of the simplification of the PURE LISP THEOREM PROVER in § 6.2.1), but now with an equation between *two non-ground* terms.

The most important case of fertilization is called "*cross-fertilization*". It occurs very often in step cases of induction proofs of equational theorems, and we have seen it already in Example 4.4 of § 4.8.1.

Neither Boyer nor Moore ever explicitly explained why cross-fertilization is "cross", but in [Moore, 1973, p. 142] we read:

> "When two equalities are involved and the fertilization was right-side" [of the induction hypothesis put] "into left-side" [of the induction conclusion,] "or left-side into right-side, it is called 'cross-fertilization'."

"Cross-fertilization" is actually a term from genetics referring to the alignment of haploid genetic code from male and female to a diploid code in the egg cell. This

---

[118] Actually, this ground term (i.e. a term without variables) here is always a *constructor* ground term (i.e. a term built-up exclusively from constructor function symbols) because the previously applied "evaluation" procedure has reduced any ground term to a constructor ground term, provided that the termination *proviso* is satisfied.

[119] Note, however, that further contextual rewriting with equalities is applied in a later stage of the Boyer–Moore waterfall, named *cross-fertilization*.

[120] See, however, Note 117 and the discussion of the PURE LISP THEOREM PROVER in § 6.3.2.

image may help to recall that only that side (i.e. left- or right-hand side of the equation) of the induction conclusion which was activated by a successful simplification is further rewritten during cross-fertilization, namely *everywhere where the same side of the induction hypothesis occurs as a redex* — just like two haploid chromosomes have to start at the same (activated) sides for successful recombination. In [Moore, 1973, p. 139] we find the reason for this: cross-fertilization frequently produces a new goal that is easy to prove because its uniform "genre" in the sense that its subterms uniformly come from just one side of the original equality.

Furthermore — for getting a sufficiently powerful new induction hypothesis in a follow-up induction — it is crucial to delete the equation used for rewriting (i.e. the old induction hypothesis), which can be remembered by the fact that — in the image — only one (diploid) genetic code remains.

The only noteworthy difference between cross-fertilization in the PURE LISP THEOREM PROVER and later Boyer–Moore theorem provers is that the generalization that consists in the deletion of the used-up equations is done in a halfhearted way: the resulting formula is equipped with a link to the deleted equation.

6.2.4 Generalization in the PURE LISP THEOREM PROVER

Generalization in the PURE LISP THEOREM PROVER works as described in § 4.9. The only difference to our presentation there is the following: Instead of just replacing all occurrences of a non-variable subterm t with a new variable z, the definition of the top function symbol of t is used to generate the definition of a new predicate p, such that $p(t)$ holds. Then the generalization of $T[t]$ becomes $T[z] \Leftarrow p(z)$ instead of just $T[z]$. The version of this automated function synthesis actually implemented in the PURE LISP THEOREM PROVER is just able to generate simple type properties, such as being a number or being a Boolean value.[121]

Note that generalization is essential for the PURE LISP THEOREM PROVER because it does not use lemmas, and so it cannot build up a more and more complex theory successively. It is clear that this limits the complexity of the theorems it can prove, because a proof can only be successful if the implemented non-backtracking heuristics work out all the way from the theorem down to the most elementary theory.

6.2.5 Elimination of Irrelevance in the PURE LISP THEOREM PROVER

There is no such stage in the PURE LISP THEOREM PROVER.

---

[121]See § 3.7 of [Moore, 1973]. As explained on Page 156f. of [Moore, 1973], Boyer and Moore failed with the trial to improve the implemented version of the function synthesis, so that it could generate a predicate on a list being ordered from a simple sorting-function.

6.2.6 Induction in the PURE LISP THEOREM PROVER

This stage of the PURE LISP THEOREM PROVER applies the induction rule of explicit induction as described in §4.8. Induction is tried only after the goal formula has been maximally simplified and generalized by repeated trips through the waterfall. The induction heuristic takes a formula as input and returns a conjunction of base and step cases to which the input formula reduces. Contrary to later Boyer–Moore theorem provers that gather the relevant information via induction schemes gleaned by preprocessing recursive definitions,[122] the induction rule of the PURE LISP THEOREM PROVER is based solely on the information provided by "evaluation" as described in §6.2.1.

Instead of trying to describe the general procedure, let us just put the induction rule of the PURE LISP THEOREM PROVER to test with two paradigmatic examples. In these examples we ignore the here irrelevant fact that the PURE LISP THEOREM PROVER actually uses a list representation for the natural numbers. The only effect of this is that the destructor p takes over the rôle of the destructor CDR.

Example 6.1 (Induction Rule in the Explicit Induction Proof of (ack4))
Let us see how the induction rule of the PURE LISP THEOREM PROVER proceeds w.r.t. the proof of (ack4) that we have seen in Example 4.5 of §4.9. The substitutions ξ_1, ξ_2 computed as instances for the induction conclusion in Example 5.5 of §5.8 suggest an overall case analysis with a base case given by $\{x \mapsto 0\}$, and two step cases given by $\xi_1 = \{x \mapsto \mathsf{s}(x'),\ y \mapsto 0\}$ and $\xi_2 = \{x \mapsto \mathsf{s}(x'),\ y \mapsto \mathsf{s}(y')\}$. The PURE LISP THEOREM PROVER requires the axioms (ack1), (ack2), (ack3) to be in destructor instead of constructor style:

(ack1') $\quad$ $\mathsf{ack}(x,y) = \mathsf{s}(y)$ $\qquad\qquad\qquad\Leftarrow\ x = 0$
(ack2') $\quad$ $\mathsf{ack}(x,y) = \mathsf{ack}(\mathsf{p}(x), \mathsf{s}(0))$ $\qquad\Leftarrow\ x \neq 0\ \wedge\ y = 0$
(ack3') $\quad$ $\mathsf{ack}(x,y) = \mathsf{ack}(\mathsf{p}(x), \mathsf{ack}(x, \mathsf{p}(y)))\ \Leftarrow\ x \neq 0\ \wedge\ y \neq 0$

"Evaluation" does not rewrite the input conjecture with this definition, but writes a "fault description" for the permanent occurrences of p as arguments of the three occurrences of ack on the right-hand sides, essentially consisting of the following three "pockets": $(\mathsf{p}(x))$, $(\mathsf{p}(x), \mathsf{p}(y))$, and $(\mathsf{p}(y))$, respectively. Similarly, the pockets gained from the fault descriptions of rewriting the input conjecture with the definition of lessp essentially consists of the pocket $(\mathsf{p}(y), \mathsf{p}(\mathsf{ack}(x,y)))$. Similar to the non-applicability of the induction template for lessp in Example 5.4 of §5.7, this fault description does not suggest any induction because one of the arguments of p in one of the pockets is not a variable. As this is not the case for the previous

---
[122] Cf. §5.8.

fault description, it suggests the set of all arguments of p in all pockets as induction variables. As this is the only suggestion, no merging of suggested inductions is applicable here.

So the PURE LISP THEOREM PROVER picks the right set of induction variables. Nevertheless, it fails to generate appropriate base and step cases, because the overall case analysis results in two base cases given by $\{x \mapsto 0\}$ and $\{y \mapsto 0\}$, and a step case given by $\{x \mapsto \mathsf{s}(x'),\ y \mapsto \mathsf{s}(y')\}$.[123] This turns the first step case of the proof of Example 4.5 into a base case. The PURE LISP THEOREM PROVER finally fails (contrary to all other Boyer–Moore theorem provers, see Examples 4.5, 5.5, and 6.11) with the step case it actually generates:

$$\mathsf{lessp}(\mathsf{s}(y'), \mathsf{ack}(\mathsf{s}(x'), \mathsf{s}(y'))) = \mathsf{true} \ \Leftarrow \ \mathsf{lessp}(y', \mathsf{ack}(x', y')) = \mathsf{true}.$$

This step case has only one hypothesis, which is neither of the two we need. □

Example 6.2 (Proof of (lessp7) by Explicit Induction with Merging)
Let us write $T(x, y, z)$ for (lessp7) of §4.4. From the proof of (lessp7) in Example 4.3 of §4.7 we can learn the following: The proof becomes simpler when we take $T(0, \mathsf{s}(y'), \mathsf{s}(z'))$ as base case (besides say $T(x, y, 0)$ and $T(x, 0, \mathsf{s}(z'))$), instead of any of $T(0, y, \mathsf{s}(z'))$, $T(0, \mathsf{s}(y'), z)$, $T(0, y, z)$. The crucial lesson from Example 4.3, however, is that the step case of explicit induction has to be

$$T(\mathsf{s}(x'), \mathsf{s}(y'), \mathsf{s}(z')) \ \Leftarrow \ T(x', y', z').$$

Note that the Boyer–Moore heuristics for using the induction rule of explicit induction look only one rewrite step ahead, separately for each occurrence of a recursive function in the conjecture.

This means that there is no way for their heuristic to apply case distinctions on variables step by step, most interesting first, until finally we end up with an instance of the induction hypothesis as in Example 4.3.

Nevertheless, even the PURE LISP THEOREM PROVER manages the pretty hard task of suggesting exactly the right step case. It requires all axioms to be in destructor style, so instead of (lessp1), (lessp2), (lessp3), we have to take:

(lessp1′) $\mathsf{lessp}(x, y) = \mathsf{false}$ $\Leftarrow y = 0$
(lessp2′) $\mathsf{lessp}(x, y) = \mathsf{true}$ $\Leftarrow y \neq 0 \ \wedge \ x = 0$
(lessp3′) $\mathsf{lessp}(x, y) = \mathsf{lessp}(\mathsf{p}(x), \mathsf{p}(y))$ $\Leftarrow y \neq 0 \ \wedge \ x \neq 0$

"Evaluation" does not rewrite any of the occurrences of lessp in the input conjecture with this definition, but writes one "fault description" for each of these occurrences about the permanent occurrences of p as argument of the one occurrence

---

[123] We can see this from a similar case on Page 164 and from the explicit description on the bottom of Page 166 in [Moore, 1973].

of lessp on the right-hand sides, resulting in one "pocket" in each fault description, which essentially consist of $((\mathsf{p}(z)))$, $((\mathsf{p}(x), \mathsf{p}(y)))$, and $((\mathsf{p}(y), \mathsf{p}(z)))$, respectively. The PURE LISP THEOREM PROVER merges these three fault descriptions to the single one $((\mathsf{p}(x), \mathsf{p}(y), \mathsf{p}(z)))$, and so suggests the proper step case indeed, although it suggests the base case $T(0, y, z)$ instead of $T(0, \mathsf{s}(y'), \mathsf{s}(z'))$, which requires some extra work, but does not result in a failure. □

6.2.7 Conclusion on the PURE LISP THEOREM PROVER

The PURE LISP THEOREM PROVER establishes the historic breakthrough regarding the heuristic automation of inductive theorem proving in theories specified by recursive function definitions.

Moreover, it is the first implementation of a prover for explicit induction going beyond most simple structural inductions over s and 0.

Furthermore, the PURE LISP THEOREM PROVER has most of the stages of the Boyer–Moore waterfall (cf. Figure 1), and these stages occur in the final order and with the final overall behavior of throwing the formulas back to the center pool after a stage was successful in changing them.

As we have seen in Example 6.1 of § 6.2.6, the main weakness of the PURE LISP THEOREM PROVER is the realization of its induction rule, which ignores most of the structure of the recursive calls in the right-hand sides of recursive function definitions.[124] In the PURE LISP THEOREM PROVER, all information on this structure that is taken into account by the induction rule comes from the fault descriptions of previous applications of "evaluation", which store only a small part of the information that is actually required for finding the proper instances for the eager instantiation of induction hypotheses required in explicit induction.

As a consequence, all induction hypotheses and conclusions of the PURE LISP THEOREM PROVER are instantiations of the input formula with mere constructor terms. Nevertheless, the PURE LISP THEOREM PROVER can generate multiple hypotheses for astonishingly complicated step cases, which go far beyond the simple ones typical for structural induction over s and 0.

Although the induction stage of the PURE LISP THEOREM PROVER is pretty underdeveloped compared to the sophisticated *recursion analysis* of the later Boyer–Moore theorem provers, it somehow contains all essential later ideas in a rudimentary form, such as recursion analysis and the merging of step cases. As we have seen in

---

[124]There are indications that the induction rule of the PURE LISP THEOREM PROVER had to be implemented in a hurry. For instance, on top of Page 168 of [Moore, 1973], we read on the PURE LISP THEOREM PROVER: "The case for n term induction is much more complicated, and is not handled in its full generality by the program."

Example 6.2, the simple merging procedure of the PURE LISP THEOREM PROVER is surprisingly successful.

The PURE LISP THEOREM PROVER cannot succeed, however, in the rare cases where a step case has to follow a destructor different from CAR and CDR (such as delfirst in § 4.5), or in the more general case that the arguments of the recursive calls contain recursively defined functions at the measured positions (such as the Ackermann function in Example 6.1).

The weaknesses and provisional procedures of the PURE LISP THEOREM PROVER we have documented, help to decompose the leap from nothing to THM, and so fulfill our historiographical intention expressed at the beginning of § 6.2.

Especially the link between symbolic evaluation and the induction rule of explicit induction described at the end of the sub-section on *"evaluation"* in § 6.2.1 (right before the sub-section on *"normalization"*) may be crucial for the success of the entire development of recursion analysis and explicit induction.

6.3 THM

"THM" is the name used in this article for the release of the prover described in [Boyer & Moore, 1979]. Note that the clearness, precision, and detail of the natural-language descriptions of heuristics in [Boyer & Moore, 1979] is probably unique.[125] To the best of our knowledge, there is no similarly broad treatment of heuristics in theorem proving, at least not in subsequent publications about Boyer–Moore theorem provers.

Except for ACL2, Boyer and Moore never gave names to their theorem provers.[126] The names "THM" (for "theorem prover"), "QTHM" ("quantified THM"), and "NQTHM" ("new quantified THM") were actually the directory names under which the different versions of their theorem provers were developed and maintained.[127] QTHM was never released and its development was discontinued soon

---

[125] In [Boyer & Moore, 1988b, p. xi] and [Boyer & Moore, 1998, p. xv] we can read about the book [Boyer & Moore, 1979]:

"The main purpose of the book was to describe in detail how the theorem prover worked, its organization, proof techniques, heuristics, etc. One measure of the success of the book is that we know of three independent successful efforts to construct the theorem prover from the book."

[126] The only further exception seems to be [Moore, 1975a, p.1], where the PURE LISP THEOREM PROVER is called "the Boyer–Moore Pure LISP Theorem Prover", because Moore wanted to stress that, though Boyer appears in the references of [Moore, 1975a] only in [Boyer & Moore, 1975], Boyer has had an equal share in contributing to the PURE LISP THEOREM PROVER right from the start.

[127] Cf. [Boyer, 2012].

after the "quantification" in NQTHM had turned out to be superior; so the name "QTHM" was never used in public. Until today, it seems that "THM" appeared in publication only as a mode in NQTHM,[128] which simulates the release previous to the release of NQTHM (i.e. before "quantification" was introduced) with a logic that is a further development of the one described in [Boyer & Moore, 1979]. It was Matt Kaufmann (*1952) who started calling the prover "NQTHM", in the second half of the 1980s.[129] The name "NQTHM" appeared for the first time in publication in [Boyer & Moore, 1988b], namely as the name of a mode in NQTHM.

In this section we describe the enormous heuristic improvements documented in [Boyer & Moore, 1979] as compared to [Moore, 1973] (cf. § 6.2). In case of the minor differences of the logic described in [Boyer & Moore, 1979] and of the later released version that is simulated by the THM mode in NQTHM as documented in [Boyer & Moore, 1988b; 1998], we try to follow the later descriptions, partly because of their elegance, partly because NQTHM is still an available program. Thus, we have entitled this section "THM" instead of "The standard reference on the Boyer–Moore heuristics [Boyer & Moore, 1979]".

From 1973 to 1981 Boyer and Moore were researchers at Xerox Palo Alto Research Center (Moore only) and — just a few miles away — at SRI International in Menlo Park (CA). From 1981 they were both professors at The University of Texas at Austin or scientists at Computational Logic Inc. in Austin (TX). So they could easily meet and work together. And — just like the PURE LISP THEOREM PROVER — the provers THM and NQTHM were again developed and implemented exclusively by Boyer and Moore.[130]

In the six years separating THM from the PURE LISP THEOREM PROVER, Boyer and Moore extended the system in four important ways that especially affect inductive theorem proving. The first major extension is the provision for an arbitrary number of inductive data types, where the PURE LISP THEOREM PROVER supported only CONS. The second is the formal provision of a definition principle with its explicit termination analysis based on well-founded relations which we discussed

---

[128]For the occurrences of "THM" in publications, and for the exact differences between the THM and NQTHM modes and logics, see Pages 256–257 and 308 in [Boyer & Moore, 1988b], as well as Pages 303–305, 326, 357, and 386 in the second edition [Boyer & Moore, 1998].

[129]Cf. [Boyer, 2012].

[130]In both [Boyer & Moore, 1988b, p. xv] and [Boyer & Moore, 1998, p. xix] we read:
"Notwithstanding the contributions of all our friends and supporters, we would like to make clear that ours is a very large and complicated system that was written entirely by the two of us. Not a single line of LISP in our system was written by a third party. Consequently, every bug in it is ours alone. Soundness is the most important property of a theorem prover, and we urge any user who finds such a bug to report it to us at once."

in § 5.5. The third major extension is the expansion of the proof techniques used by the waterfall, notably including the use of previously proved theorems, most often as rewrite rules via what would come to be called "contextual rewriting", and by which the THM user can "guide" the prover by posing lemmas that the system cannot discover on its own. The fourth major extension is the synthesis of induction schemes from definition-time termination analysis and the application and manipulation of those schemes at proof time to create "appropriate" inductions for a given formula, in place of the PURE LISP THEOREM PROVER's less structured reliance on symbolic evaluation. We discuss THM's inductive data types, waterfall, and induction schemes below.

By means of the new *shell principle*,[131] it is now possible to define new data types by describing the *shell*, a constructor with at least one argument, each of whose arguments may have a simple type restriction, and the optional *base object*, a nullary constructor.[132] Each argument of the shell can be accessed[133] by its destructor, for which a name and a default value (for the sake of totality) have to be given in addition. The user also has to supply a name for the predicate that recognizes[134] the objects of the new data type (as the logic remains untyped).

NIL lost its elementary status and is now an element of the shell PACK of symbols.[134] T and F now abbreviate the nullary function calls (TRUE) and (FALSE), respectively, which are the only Boolean values. Any argument with Boolean intention besides F is taken to be T (including NIL).

---

[131] Cf. [Boyer & Moore, 1979, p. 37ff.].

[132] Note that this restriction to at most two constructors, including exactly one with arguments, is pretty uncomfortable. For instance, it neither admits simple enumeration types (such as the Boolean values), nor disjoint unions (e.g., as part of the popular record types with variants, say of [Wirth, 1971]). Moreover, mutually recursive data types are not possible, such as and-or-trees, where each element is a list of or-and-trees, and vice versa, as given by the following four constructors:
empty-or-tree : or-tree;   or : and-tree, or-tree → or-tree;
empty-and-tree : and-tree;   and : or-tree, and-tree → and-tree.

[133] Actually, in the jargon of [Boyer & Moore, 1979; 1988b; 1998], the destructors are called *accessor functions*, and the type predicates are called *recognizer functions*.

[134] There are the following two different declarations for the shell PACK:
In [Boyer & Moore, 1979], the shell CONS is defined after the shell PACK because NIL is the default value for the destructors CAR and CDR; moreover, NIL is an abbreviation for (NIL), which is the base object of the shell PACK.
In [Boyer & Moore, 1988b; 1998], however, the shell PACK is defined after the shell CONS, we have (CAR NIL) = 0, the shell PACK has no base object, and NIL just abbreviates
(PACK (CONS 78 (CONS 73 (CONS 76 0)))).
When we discuss the logic of [Boyer & Moore, 1979], we tacitly use the shells CONS and PACK as described in [Boyer & Moore, 1988b; 1998].

Instead of discussing the shell principle in detail with all its intricacies resulting from the untyped framework, we just present the first two shells:

1. The shell (ADD1 X1) of the *natural numbers*, with
   - type restriction (NUMBERP X1),
   - base object (ZERO), abbreviated by 0,
   - destructor[134] SUB1 with default value 0, and
   - type predicate[134] NUMBERP.

2. The shell (CONS X1 X2) of *pairs*, with
   - destructors CAR with default value 0,
     CDR with default value 0, and
   - type predicate LISTP.

According to the shell principle, these two shell declarations add axioms to the theory, which are equivalent to the following ones:

#	Axioms Generated by Shell ADD1	Axioms Generated by Shell CONS
0.1	(NUMBERP X) = T $\vee$ (NUMBERP X) = F	(LISTP X) = T $\vee$ (LISTP X) = F
0.2	(NUMBERP (ADD1 X1)) = T	(LISTP (CONS X1 X2)) = T
0.3	(NUMBERP 0) = T	
0.4	(NUMBERP T) = F	(LISTP T) = F
0.5	(NUMBERP F) = F	(LISTP F) = F
0.6		(LISTP X) = F $\vee$ (NUMBERP X) = F
1	(ADD1 (SUB1 X)) = X $\Leftarrow$ X $\neq$ 0 $\wedge$ (NUMBERP X) = T	(CONS (CAR X) (CDR X)) = X $\Leftarrow$ (LISTP X) = T
2	(ADD1 X1) $\neq$ 0	
3	(SUB1 (ADD1 X1)) = X1 $\Leftarrow$ (NUMBERP X1) = T	(CAR (CONS X1 X2)) = X1 (CDR (CONS X1 X2)) = X2
4	(SUB1 0) = 0	
5.1	(SUB1 X) = 0 $\Leftarrow$ (NUMBERP X) = F	(CAR X) = 0 $\Leftarrow$ (LISTP X) = F (CDR X) = 0 $\Leftarrow$ (LISTP X) = F
5.2	(SUB1 (ADD1 X1)) = 0 $\Leftarrow$ (NUMBERP X1) = F	
L1 [135]	(ADD1 X) = (ADD1 0) $\Leftarrow$ (NUMBERP X) = F	
L2 [136]	(NUMBERP (SUB1 X)) = T	

---

[135] Proof of Lemma L1 from 0.2, 1–2, 5.2: Under the assumption of (NUMBERP X) = F, we show (ADD1 X) = (ADD1 (SUB1 (ADD1 X))) = (ADD1 0). The first step is a backward application of the conditional equation 1 via {X $\mapsto$ (ADD1 X)}, where the condition is fulfilled because of 2 and 0.2. The second step is an application of 5.2, where the condition is fulfilled by assumption.

Note that the two occurrences of "(NUMBERP X1)" in Axioms 3 and 5.2 are exactly the ones that result from the type restriction of ADD1. Moreover, the occurrence of "(NUMBERP X)" in Axiom 0.6 is allocated at the right-hand side because the shell ADD1 is declared *before* the shell CONS.

Let us discuss the axioms generated by declaration of the shell ADD1. Roughly speaking, Axioms 0.1–0.3 are return-type declarations, Axioms 0.4–0.6 are about disjointness of types, Axiom 1 and Lemma L2 imply the axiom (nat1) from § 4.4, Axioms 2 and 3 imply axioms (nat2) and (nat3), respectively. Axioms 4 and 5.1–5.2 overspecify SUB1. Note that Lemma L1 is equivalent to 5.2 under 0.2–0.3 and 1–3.

Analogous to Lemma L1, every shell forces each argument not satisfying its type restriction into behaving like the default object of the argument's destructor.

By contrast, the arguments of the shell CONS (just as every shell argument without type restriction) are not forced like this, and so — a clear advantage of the untyped framework — even objects of later defined shells (such as PACK) can be properly paired by the shell CONS. For instance, although NIL belongs to the shell PACK defined after the shell CONS (and so (CDR NIL) = 0),[135] we have (CAR (CONS NIL NIL)) = NIL by Axiom 3.

Nevertheless, the shell principle also allows us to declare a shell

(CONSNAT X1 X2)

of the *lists of natural numbers only* — similar to the ones of § 4.5 — say, with a type predicate LISTNATP, type restrictions (NUMBERP X1), (LISTNATP X2), base object (NILNAT), and destructors CARNAT, CDRNAT with default values 0, (NILNAT), respectively.

Let us now come to the admissible definitions of new functions in THM. In § 5 we have already discussed the *definition principle*[137] of THM in detail. The definition of recursive functions has not changed compared to the PURE LISP THEOREM PROVER besides that a function definition is admissible now only after a termination proof, which proceeds as explained in § 5.5. To this end, THM can apply its additional axiom of the well-foundedness of the irreflexive ordering LESSP on the natural numbers,[138] and the theorem of the well-foundedness of the lexicographic combination of two well-founded orderings.

---

[136] Proof of Lemma L2 from 0.1–0.3, 1–4, 5.1–5.2 by *argumentum ad absurdum*:
For a counterexample X, we get (SUB1 X) $\neq$ 0 by 0.3, as well as (NUMBERP (SUB1 X)) = F by 0.1. From the first we get X $\neq$ 0 by 4, and (NUMBERP X) = T by 5.1 and 0.1. Now we get the contradiction (SUB1 X) = (SUB1 (ADD1 (SUB1 X))) = (SUB1 (ADD1 0)) = 0; the first step is a backward application of the conditional equation 1, the second of L1, and the last of 3 (using 0.3).

[137] Cf. [Boyer & Moore, 1979, p. 44f.].

6.3.1 Simplification in THM

Just as in § 6.2, we will now again follow the Boyer–Moore waterfall (cf. Figure 1) and sketch how the stages of the waterfall are realized in THM in comparison to the PURE LISP THEOREM PROVER.

We discussed simplification in the PURE LISP THEOREM PROVER in § 6.2.1. Simplification in THM is covered in Chapters VI–IX of [Boyer & Moore, 1979], and the reader interested in the details is strongly encouraged to read these descriptions of heuristic procedures for simplification.

To compensate for the extra complication of the untyped approach in THM, which has a much higher number of interesting soft types than the PURE LISP THEOREM PROVER, soft-typing rules are computed for each new function symbol based on types that are disjunctions (actually: bit-vectors) of the following disjoint types: one for T, one for F, one for each shell, and one for objects not belonging to any of these.[139] These soft-typing rules are pervasively applied in all stages of the theorem prover, which we cannot discuss here in detail. Some of these rules can be expressed in the LISP logic language as a theorem and presented in this form to the human users. Let us see two examples on this.

Example 6.3 *(continuing Example 5.1 of § 5.2)*
As THM knows (NUMBERP (FIX X)) and (NUMBERP (ADD1 X)), it produces the theorem (NUMBERP (PLUS X Y)) immediately after the termination proof for the definition of PLUS in Example 5.1. Note that this would neither hold in case of

---

[138] See Page 52f. of [Boyer & Moore, 1979] for the informal statement of this axiom on well-foundedness of LESSP.

Because THM is able to prove (LESSP X (ADD1 X)), well-foundedness of LESSP would imply — together with Axiom 1 and Lemma L2 — that THM admits only the standard model of the natural numbers, as explained in Note 43.

Matt Kaufmann, however, was so kind and made clear in a private e-mail communication that non-standard models are not excluded, because the statement "We assume LESSP to be a well-founded relation." of [Boyer & Moore, 1979, p. 53] is actually to be read as the well-foundedness of the formal definition of § 4.1 with the *additional assumption* that the predicate Q must be definable in THM.

Note that in Pieri's argument on the exclusion of non-standard models (as described in Note 43), it is not possible to replace the application of the reflexive and transitive closure of the successor relation s (which is not definable in first-order logic) to an arbitrary natural number x with the THM-definable predicate
$$\{ \ Y \ | \ (\text{NUMBERP Y}) = T \ \wedge \ ((\text{LESSP Y } x) = T \ \vee \ Y = x) \ \},$$
because (by the THM-analogue of axiom (lessp2′) of Example 6.2 in § 6.2.6) this predicate will contain 0 as a minimal element even for a non-standard natural number x; thus, in non-standard models, LESSP is a *proper* super-relation of the reflexive and transitive closure of s.

[139] See Chapter VI in [Boyer & Moore, 1979].

non-termination of PLUS, nor if there were a simple Y instead of (FIX Y) in the definition of PLUS. In the latter case, THM would only register that the return-type of PLUS is among NUMBERP and the types of its second argument Y. □

Example 6.4 As THM knows that the type of APPEND is among LISTP and the type of its second argument, it produces the theorem (LISTP (FLATTEN X)) immediately after the termination proof for the following definition:
  (FLATTEN X) = (IF (LISTP X)
                    (APPEND (FLATTEN (CAR X)) (FLATTEN (CDR X)))
                    (CONS X NIL))   □

The standard representation of a propositional expression has improved from the multifarious LISP representation of the PURE LISP THEOREM PROVER toward today's standard of clausal representation. A *clause* is a disjunctive list of literals. *Literals*, however, deviating from the standard of being optionally negated atoms, are just LISP terms here, because every LISP function can be seen as a predicate.

This means that the "water" of the waterfall now consists of clauses, and the conjunction of all clauses in the waterfall represents the proof task.

Based on this clausal representation, we find a full-fledged description of *contextual rewriting* in Chapter IX of [Boyer & Moore, 1979], and its applications in Chapters VII–IX. This description comes some years before the term "contextual rewriting" became popular in automated theorem proving, and the term does not appear in [Boyer & Moore, 1979]. This may be the first description of contextual rewriting in the history of logic, unless one counts the rudimentary contextual rewriting in the "reduction" of the PURE LISP THEOREM PROVER as such.[140]

As indicated before, the essential idea of contextual rewriting is the following: While focusing on one literal of a clause for simplification, we can assume all other literals — the *context* — to be false, simply because the literal in focus is irrelevant otherwise. Especially useful are literals that are negated equations, because they can be used as a ground term-rewrite system. A non-equational literal t can always be taken to be the negated equation $(t \neq F)$. The free universal variables of a clause have to be treated as constants during contextual rewriting.[141]

---

[140]Cf. § 6.2.1.

[141]This has the advantage that we could take any well-founded ordering that is total on ground terms and run the terminating ground version of a Knuth–Bendix completion procedure [Knuth & Bendix, 1970] for all literals in a clause representation that have the form $l_i \neq r_i$, and replace the literals of this form with the resulting confluent and terminating rewrite system and normalize the other literals of the clause with it. Note that this transforms a clause into an equivalent one. None of the Boyer–Moore theorem provers does this, however.

To bring contextual rewriting to full power, all occurrences of the function symbol IF in the literals of a clause are expelled from the literals as follows. If the condition of an IF-expression can be simplified to be definitely false F or definitely true (i.e. non-F, e.g. if F is not set in the bit-vector as a potential type), then the IF-expression is replaced with its respective case. Otherwise, after the IF-expression could not be removed by those rewrite rules for IF whose soundness depends on termination,[142] it is moved to the top position (outside-in), by replacing each case with itself in the IF's context, such that the literal $C[(\text{IF } t_0\ t_1\ t_2)]$ is intermediately replaced with (IF t_0 $C[t_1]$ $C[t_2]$), and then this literal splits its clause in two: one with the two literals (NOT t_0) and $C[t_1]$ in place of the old one, and one with t_0 and $C[t_2]$ instead.

THM eagerly removes variables in solved form: If the variable X does not occur in the term t, but the literal $(\text{X} \neq t)$ occurs in a clause, then we can remove that literal after rewriting all occurrences of X in the clause to t. This removal is an equivalence transformation, because the single remaining occurrence of X is implicitly universally quantified and so $(\text{X} \neq t)$ must be false because it implies $(t \neq t)$. Alternatively, the removal can be seen as a resolution step with the axiom of reflexivity.

It now remains to describe the rewriting with function definitions and with lemmas tagged for rewriting, where the context of the clause is involved again.

Non-recursive function definitions are always unfolded by THM.

Recursive function definitions are treated in a way very similar to that of the PURE LISP THEOREM PROVER. The criteria on the unfolding of a function call of a recursively defined function f still depend solely on the terms introduced as arguments in the recursive calls of f in the body of f, which are accessed during the simplification of the body. But now, instead of rejecting the unfolding in case of the presence of new destructor terms in the simplified recursive calls, rejections are based on whether the simplified recursive calls contain subterms not occurring elsewhere in the clause. That is, an unfolding is approved if all subterms of the simplified recursive calls already occur in the clause. This basic *occurrence heuristic* is one of the keys to THM's success at induction. As we will see, instead of the PURE LISP THEOREM PROVER's phrasing of inductive arguments with "constructors in the conclusion", such as $P(\mathsf{s}(x))) \Leftarrow P(x)$, THM uses "destructors in the hypothesis", such as $(P(x) \Leftarrow P(\mathsf{p}(x))) \Leftarrow x \neq 0$. Thanks to the occurrence heuristic, the very presence of a well-chosen induction hypothesis gives the rewriter "permission" to unfold certain recursive functions in the induction conclusion (which is possible because all function definitions are in destructor style).

---

[142] These rewrite rules whose soundness depends on termination are (IF X Y Y) = Y; (IF X X F) = X; and for Boolean X: (IF X T F) = X; tested for applicability in the given order.

There are also two less important criteria which individually suffice to unblock the unfolding of recursive function definitions:

1. An increase of the number of arguments of the function to be unfolded that are constructor ground terms.

2. A decrease of the number of function symbols in the arguments of the function to be unfolded at the measured positions of an induction template for that function.
   So the clause
   $$C[\mathsf{lessp}(x, \mathsf{s}(y))]$$
   will be expanded by (lessp2′), (lessp3′), and (p1) into the clauses
   $$x \neq 0, \ C[\mathsf{true}]$$
   and
   $$x = 0, \ C[\mathsf{lessp}(\mathsf{p}(x), y)]$$
   — even if $\mathsf{p}(x)$ is a newly occurring subterm! — because the second argument position of lessp is such a set of measured positions according to Example 6.8 of § 6.3.7.[143]

THM is able to exploit previously proved lemmas. When the user submits a theorem for proof, the user tags it with tokens indicating how it is to be used in the future *if it is proved*. THM supports four non-exclusive tags and they indicate that the lemma is to be used as a rewrite rule, as a rule to eliminate destructors, as a rule to restrict generalizations, or as a rule to suggest inductions. The paradigm of tagging theorems for use by certain proof techniques focus the user on developing general "tactics" (within a limited framework of very abstract control), while allowing the user to think mainly about relevant mathematical truths. This paradigm has been a hallmark of all Boyer–Moore theorem provers since THM and partially accounts for their reputation of being "automatic".

Rewriting with lemmas that have been proved and then tagged for rewriting — so-called *rewrite lemmas* — differs from rewriting with recursive function definitions mainly in one aspect: There is no need to block them because the user has tagged them explicitly for rewriting, and because rewrite lemmas have the form of conditional equations instead of unconditional ones. Simplification with lemmas tagged for rewriting and the heuristics behind the process are nicely described in [Schmidt-Samoa, 2006c], where a rewrite lemma is not just tagged for rewriting, but where the user can also mark the condition literals on how they should be dealt

---

[143] See Page 118f. of [Boyer & Moore, 1979] for the details of the criteria for unblocking the unfolding of function definitions.

with. In THM there is no lazy rewriting with rewrite lemmas, i.e. no case splits are introduced to be able to apply the lemma.[144] This means that all conditions of the rewrite lemma have to be shown to be fulfilled in the current context. In partial compensation there is a process of backward chaining, i.e. the conditions can be shown to be fulfilled by the application of further conditional rewrite lemmas. The termination of this backward chaining is achieved by avoiding the generation of conditions into which the previous conditions can be homeomorphically embedded.[145] In addition, rewrite lemmas can introduce IF-expressions, splitting the rewritten clause into cases. There are provisions to instantiate extra variables of conditions eagerly, which is necessary because there are no existential variables.[146]

Some collections of rewrite lemmas can cause THM's rewriter not to terminate.[147] For permutative rules like commutativity, however, termination is assured by simple term ordering heuristics.[148]

6.3.2 Destructor Elimination in THM

We have already seen constructors such as s (in THM: ADD1) and cons (CONS) with the destructors p (SUB1) and car (CAR), cdr (CDR), respectively.

Example 6.5 (From Constructor to Destructor Style and back)

We have presented several function definitions both in constructor and in destructor style. Let us do careful and generalizable equivalence transformations (reverse step justified in parentheses) starting with the constructor-style rule (ack3) of §4.4:

$\quad$ ack(s(x), s(y)) = ack(x, ack(s(x), y)).

Introduce (delete) the solved variables x' and y' for the constructor terms s(x) and s(y) occurring on the left-hand side, respectively, and add (delete) two further conditions by applying the definition (p1$'$) (cf. §4.4) twice.

---

[144] Matt Kaufmann and J Strother Moore added support for "forcing" and "case split" annotations to ACL2 in the mid-1990s.

[145] See Page 109ff. of [Boyer & Moore, 1979] for the details.

[146] See Page 111f. of [Boyer & Moore, 1979] for the details.

[147] Non-termination of rewriting caused the Boyer–Moore theorem provers to run forever or exhaust the LISP stack or heap — except ACL2, which maintains its own user-adjustable stack size and gives a coherent error on stack overflow without crashing the LISP system. NQTHM introduced special tools to track down the rewriting process via the rewrite call stack (namely BREAK-REWRITE, after setting (MAINTAIN-REWRITE-PATH T)) and to count the applications of a rewrite rule (namely ACCUMULATED-PERSISTENCE), so the problematic rules can easily be detected and the user can disable them. See §12 of [Boyer & Moore, 1988b; 1998] for the details.

[148] See Page 104f. of [Boyer & Moore, 1979] for the details.

$$\mathsf{ack}(\mathsf{s}(x),\mathsf{s}(y))=\mathsf{ack}(x,\mathsf{ack}(\mathsf{s}(x),y)) \;\Leftarrow\; \begin{pmatrix} x'=\mathsf{s}(x) \;\land\; \mathsf{p}(x')=x \\ \land\;\; y'=\mathsf{s}(y) \;\land\; \mathsf{p}(y')=y \end{pmatrix}.$$

Normalize the conclusion with leftmost equations of the condition from right to left (left to right).

$$\mathsf{ack}(x',y')=\mathsf{ack}(x,\mathsf{ack}(x',y)) \;\Leftarrow\; \begin{pmatrix} x'=\mathsf{s}(x) \;\land\; \mathsf{p}(x')=x \\ \land\;\; y'=\mathsf{s}(y) \;\land\; \mathsf{p}(y')=y \end{pmatrix}.$$

Normalize the conclusion with rightmost equations of the condition from right to left (left to right).

$$\mathsf{ack}(x',y')=\mathsf{ack}(\mathsf{p}(x'),\mathsf{ack}(x',\mathsf{p}(y'))) \;\Leftarrow\; \begin{pmatrix} x'=\mathsf{s}(x) \;\land\; \mathsf{p}(x')=x \\ \land\;\; y'=\mathsf{s}(y) \;\land\; \mathsf{p}(y')=y \end{pmatrix}.$$

Add (Delete) two conditions by applying axiom (nat2) twice.

$$\mathsf{ack}(x',y')=\mathsf{ack}(\mathsf{p}(x'),\mathsf{ack}(x',\mathsf{p}(y'))) \;\Leftarrow\; \begin{pmatrix} x'=\mathsf{s}(x) \;\land\; \mathsf{p}(x')=x \;\land\; x'\neq 0 \\ \land\;\; y'=\mathsf{s}(y) \;\land\; \mathsf{p}(y')=y \;\land\; y'\neq 0 \end{pmatrix}.$$

Delete (Introduce) the leftmost equations of the condition by applying lemma (s1′) (cf. § 4.4) twice, and delete (introduce) the solved variables x and y for the destructor terms $\mathsf{p}(x')$ and $\mathsf{p}(y')$ occurring in the left-hand side of the equation in the conclusion, respectively.

$$\mathsf{ack}(x',y')=\mathsf{ack}(\mathsf{p}(x'),\mathsf{ack}(x',\mathsf{p}(y'))) \;\Leftarrow\; x'\neq 0 \land y'\neq 0.$$

Up to renaming of the variables, this is the destructor-style rule (ack3′) of Example 6.1 (cf. § 6.2.6). □

Our data types are defined inductively over constructors.[149] Therefore constructors play the main rôle in our semantics, and practice shows that step cases of simple induction proofs work out much better with constructors than with the respective destructors, which are secondary (i.e. defined) operators in our semantics and have a more complicated case analysis in applications.

For this reason — contrary to the PURE LISP THEOREM PROVER — THM applies destructor elimination to the clauses in the waterfall, but not (as in Example 6.5) to the defining equations. This application of destructor elimination has actually two further positive effects:

1. It tends to standardize the representation of a clause in the sense that the numbers of occurrences of identical subterms tend to be increased.

---

[149] Here the term "inductive" means the following: We start with the empty set and take the smallest fixpoint under application of the constructors, which contains only finite structures, such as natural numbers and lists. Co-inductively over the destructors we would obtain different data types, because we start with the universal class and obtain the greatest fixed point under inverse application of the destructors, which typically contains infinite structures. For instance, for the unrestricted destructors car, cdr of the list of natural numbers list(nat) of § 4.5, we co-inductively obtain the data type of infinite streams of natural numbers.

2. Destructor elimination also brings the subterm property in line with the substructure property; e.g., Y is both a sub-structure of (CONS X Y) and a subterm of it, whereas (CDR Z) is a sub-structure of Z in case of (LISTP Z), but not a subterm of Z.

Both effects improve the chances that the clause passes the follow-up stages of cross-fertilization and generalization with good success.[150]

As noted earlier, the PURE LISP THEOREM PROVER does induction using step cases with constructors, such as $P(\mathsf{s}(x)) \Leftarrow P(x)$, whereas THM does induction using step cases with destructors, such as
$$(\ P(x) \ \Leftarrow \ P(\mathsf{p}(x)) \ ) \ \Leftarrow \ x \neq 0.$$
So destructor elimination was not so urgent in the PURE LISP THEOREM PROVER, simply because there were fewer destructors around. Indeed, the stage "destructor elimination" does not exist in the PURE LISP THEOREM PROVER.

THM does not do induction with constructors because there are generalized destructors that do not have a straightforward constructor (see below), and because the induction rule of explicit induction has to fix in advance whether the step cases are destructor or constructor style. So with destructor style in all step cases and in all function definitions, explicit induction and recursion in THM choose the style that is always applicable. Destructor elimination then confers the advantages of constructor-style proofs when possible.

Example 6.6 (A Generalized Destructor Without Constructor)
A generalized destructor that does not have a straightforward constructor is the function delfirst defined in § 4.5. To verify the correctness of a deletion-sort algorithm based on delfirst, a useful step case for an induction proof is of the form[151]
$$(\ P(l) \ \Leftarrow \ P(\mathsf{delfirst}(\mathsf{max}(l),l)) \ ) \ \Leftarrow \ l \neq \mathsf{nil}.$$
A constructor version of this induction scheme would need something like an insertion function with an additional free variable indicating the position of insertion — resulting in proof obligations more difficult than the ones resulting directly from the algorithm to be verified. □

Proper destructor functions take only one argument. The generalized destructor delfirst we have seen in Example 6.6 has actually two arguments; the second one is the *proper destructor argument* and the first is a *parameter*. After the elimination of a set of destructors, the terms at the parameter positions of the destructors are typically still present, whereas all the terms at the positions of the proper destructor arguments are removed.

---

[150] See Page 114ff. of [Boyer & Moore, 1979] for a nice example for the advantage of destructor elimination for cross-fertilization.

Example 6.7 (Division+Remainder: a pair of Generalized Destructors)
In case of $y \neq 0$, we can construct each natural number x in the form of $(q*y)+r$ with $\mathsf{lessp}(r,y) = \mathsf{true}$. The related generalized destructors are the quotient $\mathsf{div}(x,y)$ of x by y, and its remainder $\mathsf{rem}(x,y)$. Note that in both functions, the first argument is the proper destructor argument and the second the parameter, which must not be 0. The rôle that the definition (p1′) and the lemma (s1′) of § 4.4 play in Example 6.5 (and which the definitions (car1′), (cdr1′) and the lemma (cons1′) of § 4.5 play in the equivalence transformations between constructor and destructor style for lists) is here taken by the following lemmas on the generalized destructors div and rem and on the generalized constructor $\lambda q, r.\ ((q*y)+r)$:

(div1′) $\quad \mathsf{div}(x,y) \ = q \ \Leftarrow\ y \neq 0\ \wedge\ (q*y)+r = x\ \wedge\ \mathsf{lessp}(r,y) = \mathsf{true}$
(rem1′) $\quad \mathsf{rem}(x,y) \ = r \ \Leftarrow\ y \neq 0\ \wedge\ (q*y)+r = x\ \wedge\ \mathsf{lessp}(r,y) = \mathsf{true}$
(+9′) $\quad (q*y)+r \ = x \ \Leftarrow\ y \neq 0\ \wedge\ q = \mathsf{div}(x,y)\ \wedge\ r = \mathsf{rem}(x,y)$

If we have a clause with the literal $y=0$, in which the destructor terms $\mathsf{div}(x,y)$ or $\mathsf{rem}(x,y)$ occur, we can — just as in the of Example 6.5 (reverse direction) — introduce the new literals $\mathsf{div}(x,y) \neq q$ and $\mathsf{rem}(x,y) \neq r$ for fresh q, r, and apply lemma (+9′) to introduce the literal $x \neq (q*y)+r$. Then we can normalize (from left to right) with the first two literals and then with the third, which is deleted[152] afterward. Then all occurrences of $\mathsf{div}(x,y)$, $\mathsf{rem}(x,y)$, and x are gone.[153] □

To enable the form of elimination of generalized destructors described in Example 6.7, THM allows the user to tag lemmas of the form (s1′), (cons1′), or (+9′) as *elimination lemmas* to perform destructor elimination. In clause representation, this form is in general the following: The first literal is of the form $(t^c = x)$, where x is a variable which does not occur in the (generalized) constructor term t^c. Moreover, t^c contains some distinct variables $y_0, \ldots, y_n$, which occur only on the left-hand sides of the first literal and of the last $n+1$ literals of the clause, which are of the form $(y_0 \neq t_0^d),\ \ldots,\ (y_n \neq t_n^d)$, for distinct (generalized) destructor terms $t_0^d, \ldots, t_n^d$.[154]

---

[151] See Page 143f. of [Boyer & Moore, 1979].

[152] To delete the first two literals after normalization of their x, we would need the lemmas
$\quad \mathsf{div}((q*y)+r, y) = q,\ \mathsf{lessp}(r,y) \neq \mathsf{true}\quad$ and $\quad \mathsf{rem}((q*y)+r, y) = r,\ \mathsf{lessp}(r,y) \neq \mathsf{true}$,
whose application adds the literal $\mathsf{lessp}(r,y) \neq \mathsf{true}$ unless already present.

[153] For a nice, but non-trivial example on why proofs tend to work out much easier after this transformation, see Page 135ff. of [Boyer & Moore, 1979].

[154] THM adds one more restriction here, namely that the generalized destructor terms have to consist of a function symbol applied to a list containing exactly the variables of the clause, besides $y_0, \ldots, y_n$.

Moreover, note that THM actually does not use our flattened form of the elimination lemmas, but the one that results from replacing each y_i in the clause with t_i^d, and then removing the literal $(y_i \neq t_i^d)$. Thus, THM would accept only the non-flattened versions of our elimination lemmas, such as (s1) instead of (s1′) (cf. § 4.4), and such as (cons1) instead of (cons1′) (cf. § 4.5).

The idea of application for destructor elimination in a given clause is, of course, the following: If, for an instance of the elimination lemma, the literals not mentioned above (i.e. in the middle of the clause, such as $y \neq 0$ in $(+9')$) occur in the given clause, and if $t_0^d, \ldots, t_n^d$ occur in the given clause as subterms, then rewrite all their occurrences with $(y_0 \neq t_0^d), \ldots, (y_n \neq t_n^d)$ from right to left and then use the first literal of the elimination lemma from right to left for further normalization.[155]

After a clause enters the destructor-elimination stage of THM, its most simple destructor (actually: the one defined first) that can be eliminated is eliminated, and destructor elimination is continued until all destructor terms introduced by destructor elimination are eliminated if possible. Then, before further destructors are eliminated, the resulting clause is returned to the center pool of the waterfall. So the clause will enter the simplification stage where the (generalized) constructor introduced by destructor elimination may be replaced with a (generalized) destructor. Then the resulting clauses re-enter the destructor-elimination stage, which may result in infinite looping.

For example, destructor elimination turns the clause

$$x' = 0, \quad C[\mathsf{lessp}(\mathsf{p}(x'), x')], \quad C'[\mathsf{p}(x'), x']$$

by the elimination lemma (s1) into the clause

$$\mathsf{s}(x) = 0, \quad C[\mathsf{lessp}(x, \mathsf{s}(x))], \quad C'[x, \mathsf{s}(x)].$$

Then, in the simplification stage of the waterfall, $\mathsf{lessp}(x, \mathsf{s}(x))$ is unfolded, resulting in the clause

$$x = 0, \quad C[\mathsf{lessp}(\mathsf{p}(x), x)], \quad C'[x, \mathsf{s}(x)]$$

and another one.[156]

Looping could result from eliminating the destructor introduced by simplification (such as it is actually the case for our destructor p in the last clause). To avoid looping, before returning a clause to the center pool of the waterfall, the variables introduced by destructor elimination (such as our variable x) are marked. (Generalized) destructor terms containing marked variables are blocked for further destructor elimination. This marking is removed only when the clause reaches the induction stage of the waterfall.[157]

---

[155] If we add the last literals of the elimination lemma to the given clause, use them for contextual rewriting, and remove them only if this can be achieved safely via application of the definitions of the destructors (which may not be possible in Example 6.7, cf. Note 152), then the elimination of destructors is an equivalence transformation. Destructor elimination in THM, however, may (over-)generalize the conjecture, because these last literals are not present in the non-flattened elimination lemma of THM and its variables y_i are actually introduced in THM by generalization. Thus, instead of trying to delete the last literals of our deletion lemmas safely, THM never adds them.

[156] The latter step is given in more detail in the context of the second of the two less important criteria of § 6.3.1 for unblocking the unfolding of $\mathsf{lessp}(x, \mathsf{s}(y))$.

6.3.3 (Cross-) Fertilization in THM

This stage has already been described in § 6.2.3. There is no noticeable difference between the PURE LISP THEOREM PROVER and THM here, besides some heuristic fine tuning.[158]

6.3.4 Generalization in THM

THM adds only one new rule to the universally applicable heuristic rules for generalization on a term t mentioned in § 4.9:

"Never generalize on a destructor term t!"

This new rule makes sense in particular after the preceding stage of destructor elimination in the sense that destructors that outlast their elimination probably carry some relevant information. Another reason for not generalizing on destructor terms is that the clause will enter the center pool in case another generalization is possible, and then the destructor elimination might eliminate the destructor term more carefully than generalization would do.[159]

The main improvement of generalization in THM over the PURE LISP THEOREM PROVER, however, is the following: Suppose again that the term t is to be replaced at all its occurrences in the clause $T[t]$ with the fresh variable z. Recall that the PURE LISP THEOREM PROVER restricts the fresh variable with a predicate synthesized from the definition of the top function symbol of the replaced term. THM instead restricts the new variable in two ways. Both ways add additional literals to the clause before the term is replaced by the fresh variable:

1. Assuming all literals of the clause $T[t]$ to be false (i.e. of type F), the bit-vector describing the soft type of t is computed and if only one bit is set (say the bit expressing NUMBERP), then, for the respective type predicate, a new literal is added to the clause (such as (NOT (NUMBERP t))).

2. The user can tag certain lemmas as *generalization lemmas*; such as
$$\text{(SORTEDP (SORT X))}$$
for a sorting function SORT; and if (SORT X) matches t, the respective instance of (NOT (SORTEDP (SORT X))) is added to $T[t]$.[160] In general, for the addition of such a literal (NOT t'), a proper subterm t' of a generalization lemma must match t.[161]

---

[157] See Page 139 of [Boyer & Moore, 1979]. In general, for more sophisticated details of destructor elimination in THM, we have to refer the reader to Chapter X of [Boyer & Moore, 1979].

[158] See Page 149 of [Boyer & Moore, 1979].

[159] See Page 156f. of [Boyer & Moore, 1979].

[160] Cf. Note 121.

6.3.5 Elimination of Irrelevance in THM

THM includes another waterfall stage not in the PURE LISP THEOREM PROVER, the elimination of irrelevant literals. This is the last transformation before we come to "induction". Like generalization, this stage may turn a valid clause into an invalid one. The main reason for taking this risk is that the subsequent heuristic procedures for induction assume all literals to be relevant: irrelevant literals may suggest inappropriate induction schemes which may result in a failure of the induction proof. Moreover, if all literals seem to be irrelevant, then the goal is probably invalid and we should not do a costly induction but just fail immediately.[162]

Let us call two literals *connected* if there is a variable that occurs in both of them. Consider the partition of a clause into its equivalence classes w.r.t. the reflexive and transitive closure of connectedness. If we have more than one equivalence class in a clause, this is an alarm signal for irrelevance: if the original clause is valid, then a subclause consisting only of the literals of one of these equivalence classes must be valid as well. This is a consequence of the equivalence of $\forall x.(A \vee B)$ with $A \vee \forall x. B$, provided that x does not occur in A. Then we should remove one of the irrelevant equivalence classes after the other from the original clause. To this end, THM has two heuristic tests for irrelevance.

1. *An equivalence class of literals is irrelevant if it does not contain any properly recursive function symbol.* Based on the assumption that the previous stages of the waterfall are sufficiently powerful to prove clauses composed only of constructor functions (i.e. shells and base objects) and functions with explicit (i.e. non-recursive) definitions, the justification for this heuristic test is the following: If the clause of the equivalence class were valid, then the previous stages of the waterfall should already have established the validity of this equivalence class.

2. *An equivalence class of literals is irrelevant if it consists of only one literal and if this literal is the application of a properly recursive function to a list of distinct variables.* Based on the assumption that the soft typing rules are sufficiently powerful and that the user has not defined a tautological, but tricky

---

[161] Moreover, the literal is actually added to the generalized clause only if the top function symbol of t does no longer occur in the literal after replacing t with x. This means that, for a generalization lemma (EQUAL (FLATTEN (GOPHER X)) (FLATTEN X)), the literal
(NOT (EQUAL (FLATTEN (GOPHER t'')) (FLATTEN t'')))
is added to $T[t]$ in case of t being of the form (GOPHER t''), but not in case of t being of the form (FLATTEN t'') where the first occurrence of FLATTEN is not removed by the generalization. See Page 156f. of [Boyer & Moore, 1979] for the details.

[162] See Page 160f. of [Boyer & Moore, 1979] for a typical example of this.

predicate,[163] the justification for this heuristic test is the following: The bit-vector of this literal must contain the singleton type of F (containing only the term F, cf. § 6.3.1); otherwise the validity of the literal and the clause would have been recognized by the stage "simplification". This means that F is most probably a possible value for some combination of arguments.

6.3.6 Induction in THM as compared to the PURE LISP THEOREM PROVER

As we have seen in § 6.2.6, the *recursion analysis* in the PURE LISP THEOREM PROVER is only rudimentary. Indeed, the whole information on the body of the recursive function definitions comes out of the poor[164] feedback of the "evaluation" procedure of the simplification stage of the PURE LISP THEOREM PROVER. Roughly speaking, this information consists only in the two facts

1. that a destructor symbol occurring as an argument of the recursive function call in the body is not removed by the "evaluation" procedure in the context of the current goal and in the local environment, and

2. that it is not possible to derive that this recursive function call is unreachable in this context and environment.

In THM, however, the first part of recursion analysis is done at *definition time*, i.e. at the time the function is defined, and applied at *proof time*, i.e. at the time the induction rule produces the base and step cases. Surprisingly, there is no reachability analysis for the recursive calls in this second part of the recursion analysis in THM. While the information in item 1 is thoroughly improved as compared to the PURE LISP THEOREM PROVER, the information in item 2 is partly weaker because all recursive function calls are assumed to be reachable during recursion analysis. The overwhelming success of THM means that the heuristic decision to abandon reachability analysis in THM was appropriate.[165]

---

[163] This assumption is critical because it often occurs that updated program code contains recursive predicates that are actually trivially true, but very tricky. See § 3.2 of [Wirth, 2004] for such an example. Moreover, users sometimes supply such predicates to suggest a particular induction ordering. For example, if we want to supply the function sqrtio of § 6.3.9 to THM, then we have to provide a complete definition, typically given by setting sqrtio to be T in all other cases. Luckily, such nonsense functions will typically not occur in any proof.

[164] See the discussion in § 6.2.7 on Example 6.1 from § 6.2.6.

[165] Note that in most cases the step formula of the reachable cases works somehow in THM, as long as no better step case was canceled because of unreachable step cases, which, of course, are trivial to prove, simply because their condition is false. Moreover, note that, contrary to *descente infinie* which can get along with the first part of recursion analysis alone, the heuristics of explicit induction have to guess the induction steps eagerly, which is always a fault-prone procedure, to be corrected by additional induction proofs, as we have seen in Example 4.4 of § 4.8.1.

6.3.7 Induction Templates generated by Definition-Time Recursion Analysis

The first part of recursion analysis in THM consists of a termination analysis of every recursive function at the time of its definition. The system does not only look for one termination proof that is sufficient for the admissibility of the function definition, but — to be able to generate a plenitude of sound sets of step formulas later — actually looks through all termination proofs in a finite search space and gathers from them all information required for justifying the termination of the recursive function definition. This information will later be used to guarantee the soundness and improve the feasibility of the step cases to be generated by the induction rule.

To this end, THM constructs valid induction templates very similar[166] to our description in § 5.5. Let us approach the idea of a valid induction template with some typical examples, which are actually the templates for the constructor-style examples of § 5.5, but now for the destructor-style definitions of lessp and ack, because only destructor-style definitions are admissible in THM.

Example 6.8 (Two Induction Templates, Different Measured Positions)
For the ordering predicate lessp as defined by (lessp1′–3′) in Example 6.2 of § 6.2.6, we get two induction templates with the sets of measured positions $\{1\}$ and $\{2\}$, respectively, both for the well-founded ordering $\lambda x, y.\ (\mathsf{lessp}(x,y) = \mathsf{true})$. The first template has the weight term (1) and the relational description
$$\{\ (\ \mathsf{lessp}(x,y),\ \{\mathsf{lessp}(\mathsf{p}(x),\mathsf{p}(y))\},\ \{x \neq 0\}\ )\ \}.$$
The second one has the weight term (2) and the relational description
$$\{\ (\ \mathsf{lessp}(x,y),\ \{\mathsf{lessp}(\mathsf{p}(x),\mathsf{p}(y))\},\ \{y \neq 0\}\ )\ \}. \qquad \square$$

Example 6.9 (One Induction Template with Two Measured Positions)
For the Ackermann function ack as defined by (ack1′–3′) in Example 6.1 of § 6.2.6, we get only one appropriate induction template. The set of its measured positions is $\{1,2\}$, because of the weight function $\mathsf{cons}((1), \mathsf{cons}((2), \mathsf{nil}))$ (in THM actually: (CONS x y)) in the well-founded lexicographic ordering

---

[166]Those parts of the condition of the equation that contain the new function symbol f must be ignored in the case conditions of the induction template because the definition of the function f is admitted in THM only *after* it has passed the termination proof.

That THM ignores the governing conditions that contain the new function symbol f is described in the 2nd paragraph on Page 165 of [Boyer & Moore, 1979]. Moreover, an example for this is the definition of OCCUR on Page 166 of [Boyer & Moore, 1979].

After one successful termination proof, however, the function can be admitted in THM, and then these conditions could actually be admitted in the templates. So the actual reason why THM ignores these conditions in the templates is that it generates the templates with the help of previously proved *induction lemmas*, which, of course, cannot contain the new function yet.

$$\lambda l, k. \ (\mathsf{lexlimless}(l, k, \mathsf{s}(\mathsf{s}(\mathsf{s}(0)))) = \mathsf{true}).$$
The relational description has two elements: For the equation (ack2′) we get
$$(\ \mathsf{ack}(x,y),\ \{\mathsf{ack}(\mathsf{p}(x),\mathsf{s}(0))\},\ \{x \neq 0\}\ ),$$
and for the equation (ack3′) we get
$$(\ \mathsf{ack}(x,y),\ \{\mathsf{ack}(x,\mathsf{p}(y)),\ \mathsf{ack}(\mathsf{p}(x),\mathsf{ack}(x,\mathsf{p}(y)))\},\ \{x \neq 0,\ y \neq 0\}\ ). \quad \square$$

To find valid induction templates automatically by exhaustive search, THM allows the user to tag certain theorems as *"induction lemmas"*. An induction lemma consists of the application of a well-founded relation to two terms with the same top function symbol w, playing the rôle of the weight term; plus a condition without extra variables, which is used to generate the case conditions of the induction template. Moreover, the arguments of the application of w occurring as the second argument of the well-founded relation must be distinct variables in THM, mirroring the left-hand side of its function definitions in destructor style.

Certain induction lemmas are generated with each shell declaration. Such an induction lemma generated for the shell ADD1, which is roughly
$$(\mathtt{LESSP\ (COUNT\ (SUB1\ X))\ (COUNT\ X))} \ \Leftarrow \ (\mathtt{NOT\ (ZEROP\ X)}),$$
suffices for generating the two templates of Example 6.8. Note that COUNT, playing the rôle of w here, is a special function in THM, which is generically extended by every shell declaration in an object-oriented style for the elements of the new shell. On the natural numbers here, COUNT is the identity. On other shells, COUNT is defined similar to our function count from § 4.5.[167]

6.3.8 Proof-Time Recursion Analysis in THM

The induction rule uses the information from the induction templates as follows: For each recursive function occurring in the input formula, all *applicable* induction templates are retrieved and turned into *induction schemes* as described in § 5.8. Any induction scheme that is *subsumed* by another one is deleted after adding its hitting ratio to the one of the other. The remaining schemes are *merged* into new ones with a higher hitting ratio, and finally, after the *flawed* schemes are deleted, the scheme with the highest hitting ratio will be used by the induction rule to generate the base and step cases.

Example 6.10 (Applicable Induction Templates)
Let us consider the conjecture (ack4) from § 4.4. From the three induction templates of Examples 6.8 and 6.9, only the second one of Example 6.8 is not applicable because the second position of lessp (which is the only measured position of that template) is changeable, but filled in (ack4) by the non-variable $\mathsf{ack}(x,y)$. $\quad \square$

From the destructor-style definitions (lessp1′–3′) (cf. Example 6.2) and (ack1′–3′) (cf. Example 6.1), we have generated two induction templates applicable to
(ack4) $\mathsf{lessp}(y, \mathsf{ack}(x,y)) = \mathsf{true}$
They yield the *two* induction schemes of Example 6.11. See also Example 5.5 for the *single* induction scheme for the constructor-style definitions (lessp1–3) and (ack1–3).

Example 6.11 (Induction Schemes)
The induction template for lessp of Example 6.8 that is applicable to (ack4) according to Example 6.10 and whose relational description contains only the triple
$$(\mathsf{lessp}(x,y), \{\mathsf{lessp}(\mathsf{p}(x), \mathsf{p}(y))\}, \{x \neq 0\})$$
yields the induction scheme with position set $\{1.1\}$ (i.e. left-hand side of first literal in (ack4)); the step-case description is $\{(_{\{x,y\}}|\mathsf{id}, \{\mu_1\}, \{y \neq 0\})\}$, where $\mu_1 = \{x \mapsto x, y \mapsto \mathsf{p}(y)\}$; the set of induction variables is $\{y\}$; and the hitting ratio is $\frac{1}{2}$.

This can be seen as follows: The substitution called ξ in the discussion of §5.8 can be chosen to be the identity substitution $_{\{x,y\}}|\mathsf{id}$ on $\{x,y\}$ because the first element of the triple does not contain any constructors. This is always the case for induction templates for destructor-style definitions such as (lessp1′–3′). The substitution called σ in §5.8 (which has to match the first element of the triple to the term (ack4)/1.1, i.e. the term at the position 1.1 in (ack4)) is
$$\sigma = \{x \mapsto y, y \mapsto \mathsf{ack}(x,y)\}.$$
So the constraints for μ_1 (which tries to match (ack4)/1.1 to the σ-instance of the second element of the triple) are: $y\mu_1 = \mathsf{p}(y)$ for the first (measured) position of lessp; and $\mathsf{ack}(x,y)\mu_1 = \mathsf{p}(\mathsf{ack}(x,y))$ for the second (unmeasured) position, which cannot be achieved and is skipped. This results in a hitting ratio of only $\frac{1}{2}$. The single measured position 1 of the induction template results in the induction variable (ack4)/1.1.1 = y.

The template for ack of Example 6.9 yields an induction scheme with the position set $\{1.1.2\}$, and the set of induction variables $\{x,y\}$. The triple
$$(\mathsf{ack}(x,y), \{\mathsf{ack}(\mathsf{p}(x), \mathsf{s}(0))\}, \{x \neq 0\})$$
(generated by the equation (ack2′)) is replaced with $(_{\{x,y\}}|\mathsf{id}, \{\mu'_{1,1}\}, \{x \neq 0\})$, where $\mu'_{1,1} = \{x \mapsto \mathsf{p}(x), y \mapsto \mathsf{s}(0)\}$. The triple
$$(\mathsf{ack}(x,y), \{\mathsf{ack}(x, \mathsf{p}(y)), \mathsf{ack}(\mathsf{p}(x), \mathsf{ack}(x, \mathsf{p}(y)))\}, \{x \neq 0, \ y \neq 0\})$$
(generated by (ack3′)) is replaced with $(_{\{x,y\}}|\mathsf{id}, \{\mu'_{2,1}, \mu'_{2,2}\}, \{x \neq 0, \ y \neq 0\})$, where $\mu'_{2,1} = \{x \mapsto x, y \mapsto \mathsf{p}(y)\}$, and $\mu'_{2,2} = \{x \mapsto \mathsf{p}(x), y \mapsto \mathsf{ack}(x, \mathsf{p}(y))\}$.

This can be seen as follows: The substitution called σ in §5.8 is $_{\{x,y\}}|\mathsf{id}$ in both cases, and so the constraints for the (measured) positions are $x\mu'_{1,1} = \mathsf{p}(x)$, $y\mu'_{1,1} = \mathsf{s}(0)$; $x\mu'_{2,1} = x$, $y\mu'_{2,1} = \mathsf{p}(y)$; $x\mu'_{2,2} = \mathsf{p}(x)$, $y\mu'_{2,2} = \mathsf{ack}(x, \mathsf{p}(y))$.
As all six constraints are satisfied, and the hitting ratio is $\frac{6}{6} = 1$.   □

An induction scheme that is either *subsumed by* or *merged into* another induction scheme adds its hitting ratio and sets of positions and induction variables to those of the other's, respectively, and then it is deleted.

The most important case of subsumption are schemes that are identical except for their position sets, where — no matter which scheme is deleted — the result is the same. The more general case of proper subsumption occurs when the subsumer provides the essential structure of the subsumee, but not vice versa.

Merging and proper subsumption of schemes — seen as binary algebraic operations — are not commutative, however, because the second argument inherits the well-foundedness guarantee alone and somehow absorbs the first argument, and so the result for swapped arguments is often undefined.

More precisely, subsumption is given if the step-case description of the first induction scheme can be injectively mapped to the step-case description of the second one, such that (using the notation of §5.8 and Example 6.11), for each step case $(\mathrm{id}, \{\, \mu_j \mid j \in J \,\}, C)$ mapped to $(\mathrm{id}, \{\, \mu'_j \mid j \in J \uplus J' \,\}, C')$, we have $C \subseteq C'$, and the set of substitutions $\{\, \mu_j \mid j \in J \,\}$ can be injectively[168] mapped to $\{\, \mu'_j \mid j \in J \uplus J' \,\}$ (w.l.o.g. say μ_i to μ'_i for $i \in J$), such that, for each $j \in J$ and $x \in \mathrm{dom}(\mu_j)$: $x \in \mathrm{dom}(\mu'_j)$; $x\mu_j = x$ implies $x\mu'_j = x$; and $x\mu_j$ is a subterm of $x\mu'_j$.

Example 6.12 (Subsumption of Induction Schemes)
In Example 6.11, the induction scheme for lessp is subsumed by the induction scheme for ack, because we can map the only element of the step-case description of the former to the second element of the step-case description of latter: the case condition $\{y \neq 0\}$ is a subset of the case condition $\{x \neq 0,\ y \neq 0\}$, and we have $\mu_1 = \mu'_{2,1}$. So the former scheme is deleted and the scheme for ack is updated to have the position set $\{1.1,\ 1.1.2\}$ and the hitting ratio $\frac{3}{2}$. □

In Example 6.2 of §6.2.6 we have already seen a rudimentary, but pretty successful kind of *merging of suggested step cases* in the PURE LISP THEOREM PROVER. As THM additionally has induction schemes, it applies a more sophisticated *merging of induction schemes* instead.

---

[167] For more details on the recursion analysis a definition time in THM, see Page 180ff. of [Boyer & Moore, 1979].

[168] From a logical viewpoint, it is not clear why this second injectivity requirement is found here, just as in different (but equivalent) form in [Boyer & Moore, 1979, p. 191, 1st paragraph]. (The first injectivity requirement may prevent us from choosing an induction ordering that is too small, cf. §6.3.9.) An omission of the second requirement would just admit a term of the subsumer to have multiple subterms of the subsumee, which seems reasonable. Nevertheless, as pointed out in §6.3.9, only practical testing of the heuristics is what matters here. See also Note 169.

Two substitutions μ_1 and μ_2 are [*non-trivially*] *mergeable* if $x\mu_1 = x\mu_2$ for each $x \in \text{dom}(\mu_1) \cap \text{dom}(\mu_2)$ [and there is a $y \in \text{dom}(\mu_1) \cap \text{dom}(\mu_2)$ with $y\mu_1 \neq y$].

Two triples $(_{V_1}|\text{id}, A_1, C_1)$ and $(_{V_2}|\text{id}, A_2, C_2)$ of two step-case descriptions of two induction schemes, each with domain $V_k = \text{dom}(\mu_k)$ for all $\mu_k \in A_k$ (for $k \in \{1, 2\}$), are [*non-trivially*] *mergeable* if for each $\mu_1 \in A_1$ there is a $\mu_2 \in A_2$ such that μ_1 and μ_2 are [non-trivially] mergeable. The result of their merging is
$$\left(\ _{V_1 \cup V_2}|\text{id},\ m(A_1, A_2),\ C_1 \cup C_2\ \right),$$
where $m(A_1, A_2)$ is the set containing all substitutions $\mu_1 \cup \mu_2$ with $\mu_1 \in A_1$ and $\mu_2 \in A_2$ such that μ_1 and μ_2 are mergeable as well as all substitutions $_{V_1 \setminus V_2}|\text{id} \cup \mu_2$ with $\mu_2 \in A_2$ for which there is no substitution $\mu_1 \in A_1$ such that μ_1 and μ_2 are mergeable.

Two induction schemes are *mergeable* if the step-case description of the first induction scheme can be injectively[169] mapped to the step-case description of the second one, such that each argument and its image are non-trivially mergeable. The step-case description of the induction scheme that results from *merging the first induction scheme into the second* contains the merging of all mergeable triples of the step-case descriptions of the first and second induction scheme, respectively.

Finally, we have to describe what it means that an induction scheme is *flawed*. This simply is the case if — after merging is completed — the intersection of its induction variables with the (common) domain of the substitutions of the step-case description of another remaining induction scheme is non-empty.

If an induction scheme is flawed by another one that cannot be merged with it, this indicates that an induction on it will probably result in a permanent clash between the induction conclusion and the available induction hypotheses at some occurrences of the induction variables.[170]

Example 6.13 (Merging and Flawedness of Induction Schemes)

Let us reconsider merging in the proof of lemma (lessp7) w.r.t. the definition of lessp via (lessp1′–3′), just as we did in Example 6.2. Let us abbreviate $p = \text{true}$ with p, just as in our very first proof of lemma (lessp7) in Example 4.3, and also following the LISP style of THM. Simplification reduces (lessp7) first to the clause

(lessp7′) $\text{lessp}(x, \text{p}(z)),\ \neg\text{lessp}(x, y),\ \neg\text{lessp}(y, z),\ z = 0$

---

[169] From a logical viewpoint, it is again not clear why an injectivity requirement is found here, just as in different (but equivalent) form in [Boyer & Moore, 1979, p. 193, 1st paragraph]. An omission of the injectivity requirement would admit to define merging as a commutative associative operation. Nevertheless, as pointed out in § 6.3.9, only practical testing of the heuristics is what matters here. See also Note 168.

[170] See Page 194f. of [Boyer & Moore, 1979] for a short further discussion and a nice example.

	pos. set	ind. var.s	step-case description	hitting ratio
1	$\{1\}$	$\{x\}$	$\{\,(_{\{x,z\}}|\text{id},\ \{\mu_1\},\ \{x\neq 0\})\,\}$	1
2	$\{2\}$	$\{x\}$	$\{\,(_{\{x,y\}}|\text{id},\ \{\mu_2\},\ \{x\neq 0\})\,\}$	1
3	$\{2\}$	$\{y\}$	$\{\,(_{\{x,y\}}|\text{id},\ \{\mu_2\},\ \{y\neq 0\})\,\}$	1
4	$\{3\}$	$\{y\}$	$\{\,(_{\{y,z\}}|\text{id},\ \{\mu_3\},\ \{y\neq 0\})\,\}$	1
5	$\{3\}$	$\{z\}$	$\{\,(_{\{y,z\}}|\text{id},\ \{\mu_3\},\ \{z\neq 0\})\,\}$	1
6	$\{2\}$	$\{x,y\}$	$\{\,(_{\{x,y\}}|\text{id},\ \{\mu_2\},\ \{x\neq 0,\ y\neq 0\})\,\}$	2
7	$\{3\}$	$\{y,z\}$	$\{\,(_{\{y,z\}}|\text{id},\ \{\mu_3\},\ \{y\neq 0,\ z\neq 0\})\,\}$	2
8	$\{2,3\}$	$\{x,y,z\}$	$\{\,(_{\{x,y,z\}}|\text{id},\ \{\mu_4\},\ \{x\neq 0,\ y\neq 0,\ z\neq 0\})\,\}$	4
9	$\{1,2,3\}$	$\{x,y,z\}$	$\{\,(_{\{x,y,z\}}|\text{id},\ \{\mu_4\},\ \{x\neq 0,\ y\neq 0,\ z\neq 0\})\,\}$	5

$$\mu_1 = \{x \mapsto \mathsf{p}(x),\ z \mapsto \mathsf{p}(z)\}, \qquad \mu_2 = \{x \mapsto \mathsf{p}(x),\ y \mapsto \mathsf{p}(y)\},$$
$$\mu_3 = \{y \mapsto \mathsf{p}(y),\ z \mapsto \mathsf{p}(z)\}, \quad \text{and} \quad \mu_4 = \{x \mapsto \mathsf{p}(x),\ y \mapsto \mathsf{p}(y),\ z \mapsto \mathsf{p}(z)\}.$$

pos. = position; ind. var.s = set of induction variables.

Figure 2: The induction schemes of Example 6.13

Then the Boyer–Moore waterfall sends this clause through three rounds of reduction between destructor elimination and simplification as discussed at the end of § 6.3.2, finally returning again to (lessp7′), but now with all its variables marked as being introduced by destructor elimination, which prevents looping by blocking further destructor elimination.

Note that the marked variables refer actually to the predecessors of the values of the original lemma (lessp7′), and that these three rounds of reduction already include all that is required for the entire induction proof, such that *descente infinie* would now conclude the proof with an induction-hypothesis application. This most nicely illustrates the crucial similarity between recursion and induction, which Boyer and Moore "exploit" ... "or, rather, contrived".[171]

The proof by explicit induction in THM, however, now just starts to compute induction schemes. The two induction templates for lessp found in Example 6.8 are applicable five times, resulting in the induction schemes 1–5 in Figure 2.

From the domains of the substitutions in the step-case descriptions, it is obvious that — among schemes 1–5 — only the two pairs of schemes 2 and 3 as well as 4 and 5 are candidates for subsumption, which is not given here, however, because the case conditions of these two pairs of schemes are not subsets of each other.

Nevertheless, these pairs of schemes merge, resulting in the schemes 6 and 7, respectively, which merge again, resulting in scheme 8.

---

[171] Cf. [Boyer & Moore, 1979, p. 163, last paragraph].

Now only the schemes 1 and 8 remain. As each of them has x as an induction variable, both schemes would be flawed if they could not be merged.

It does not matter that the scheme 1 is subsumed by scheme 8 simply because the phase of subsumption is already over; but they are also mergeable, actually with the same result as subsumption would have, namely the scheme 9, which admits us to prove the generic step-case formula it describes without further induction, and so THM achieves the crucial task of heuristic anticipation of an appropriate induction hypotheses, just as well as the PURE LISP THEOREM PROVER.[172]   □

6.3.9 Conclusion on THM

Logicians reading on THM may ask themselves many questions such as: Why is merging of induction schemes — seen as a binary algebraic operation — not realized to satisfy the constraint of associativity, so that the result of merging become independent of the order of the operations? Why does merging not admit the subterm-property in the same way as subsumption of induction schemes does? Why do some of the injectivity requirements[173] of subsumption and mergeability lack a meaningful justification, and how can it be that they do not matter?

The answer is trivial, although it is easily overlooked: The part of the automation of induction discussed here belongs more to the field of heuristics than to the field of logics. Therefore, the final judgment cannot come from logical and intellectual adequacy and comprehensibility — which are not much more applicable here than in the field of neural nets for instance — but must come from complete testing with a huge and growing corpus of example theorems. A modification of an operation, say merging of induction schemes, that may have some practical advantages for some examples or admit humans some insight or understanding, can be accepted only if it admits us to run, as efficiently as before, all the lemmas that could be automatically proved with the system before. All in all, logical and formal considerations may help us to find new heuristics, but they cannot play any rôle in their evaluation.[174]

---

[172] The base cases show no improvement to the proof with the PURE LISP THEOREM PROVER in Example 6.2 and a further additional, but also negligible overhead is the preceding reduction from (lessp7) over (lessp7′) to a version of (lessp7′) with marked variables.

[173] Cf. Notes 168 and 169.

[174] While Christoph Walther is well aware of the primacy of testing in [Walther, 1992; 1993], this awareness is not reflected in the sloppy language of the most interesting papers [Stevens, 1988] and [Bundy &al., 1989]: Heuristics cannot be "bugged" or "have serious flaws", unless this would mean that they turn out to be inferior to others w.r.t. a standard corpus. A "rational reconstruction" or a "meta-theoretic analysis" may help to guess even superior heuristics, but they may not have any epistemological value *per se*.

Moreover, it is remarkable that the well-founded relation that is expressed by the subsuming induction scheme is smaller than that expressed by the subsumed one, and the relation expressed by a merged scheme is typically smaller than those expressed by the original ones. This means that the newly generated induction schemes do not represent a more powerful induction ordering (say, in terms of Noetherian induction), but actually achieve an improvement w.r.t. the eager instantiation of the induction hypothesis (both for a direct proof and for generalization), and provide case conditions that further a successful generalization without further case analysis.

Since the end of the 1970s until today, THM has set the standard for explicit induction; moreover, THM and its successors NQTHM and ACL2 have given many researchers a hard time trying to demonstrate weaknesses of the explicit-induction heuristics, because examples carefully devised to fail with certain steps of the construction of induction schemes (or other stages of the waterfall) tend to end up with alternative proofs not imagined before.

Restricted to the mechanization of the selection of an appropriate induction scheme for explicit induction, no significant overall progress has been seen beyond THM and we do not expect any such progress for the future. A heuristic approach that has to anticipate appropriate induction steps with a lookahead of one individual rewrite step for each recursive function occurring in the input formula cannot go much further than the carefully developed and exhaustively tested explicit-induction heuristics of THM.

Working with THM (or NQTHM) for the first time will always fascinate informaticians and mathematicians, simply because it helps to save more time with the standard everyday inductive proof work than it takes, and the system often comes up with completely unexpected proofs. Mathematicians, however, should be warned that the less trivial mathematical proofs that require some creativity and would deserve to be explicated in a mathematics lecture, will require some hints, especially if the induction ordering is not a combination of the termination orderings of the given function definitions. This is already the case for the simple proofs of the lemma on the irrationality of the square root of two, simply because the induction orderings of the typical proofs exist only under the assumption that the lemma is wrong. To make THM find the standard proof, the user has to define a function such as sqrtio with a defining rule such as (sqrtio1) in Figure 3 on the following page.

Note that the condition of (sqrtio1) cannot be fulfilled. The three different occurrences of sqrtio on the right-hand side of the positive/negative-conditional equation become immediately clear from Figure 3. Actually, any single one of these occurrences is sufficient for a proof of the irrationality lemma with THM, provided that we give the hint that the induction templates of sqrtio should be used for computing the induction schemes, in spite of the fact that sqrtio does not occur in the lemma.

(sqrtio1)　　sqrtio(x, y)
　　　　　= and(sqrtio$(y, \text{div}(x, \mathsf{s}(\mathsf{s}(0))))$,
　　　　　　and(sqrtio$(\mathsf{s}(\mathsf{s}(0)) * (x-y), (\mathsf{s}(\mathsf{s}(0))) * y) - x)$,
　　　　　　sqrtio$((\mathsf{s}(\mathsf{s}(0))) * y) - x, x - y)))$
$$\Leftarrow\ x * x = \mathsf{s}(\mathsf{s}(0)) * y * y\ \wedge\ y \neq 0$$

The arguments of the three recursive calls of sqrtio in the right-hand side represent the step from the left-hand side given as the the triangle with right angle at F to those at C, G, and B, respectively.

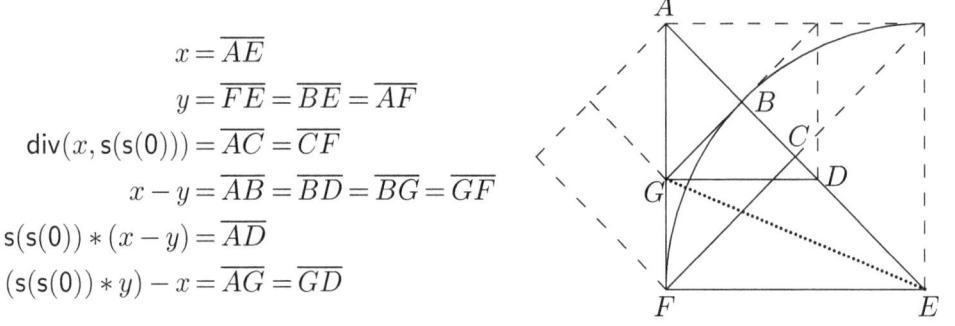

$$x = \overline{AE}$$
$$y = \overline{FE} = \overline{BE} = \overline{AF}$$
$$\text{div}(x, \mathsf{s}(\mathsf{s}(0))) = \overline{AC} = \overline{CF}$$
$$x - y = \overline{AB} = \overline{BD} = \overline{BG} = \overline{GF}$$
$$\mathsf{s}(\mathsf{s}(0)) * (x - y) = \overline{AD}$$
$$(\mathsf{s}(\mathsf{s}(0)) * y) - x = \overline{AG} = \overline{GD}$$

Figure 3: Four possibilities to descend with rational representations of $\sqrt{2}$:

6.4 NQTHM

Subsequent theorem provers by Boyer and Moore did not add much to the selection of an appropriate induction scheme. While both NQTHM and ACL2 have been very influential in theorem proving, their inductive heuristics are nearly the same as those in THM and their waterfalls have quite similar structures. As we are concerned here only with the essential history of the mechanization of induction, we just sketch most interesting developments since 1979.

The one change from THM to NQTHM that most directly affected the inductions carried out by the system is the abandonment of fixed lexicographic relations on natural numbers as the only available well-founded relations. NQTHM introduces a formal representation of the ordinals up to ε_0, i.e. up to $\omega^{\omega^{\cdot^{\cdot^{\cdot}}}}$, and assumes that the "less than" relation on such ordinals is well-founded. This did not change the induction heuristics themselves, it just allowed the admission of more complex function definitions and the justification of more sophisticated induction templates.

After the publication of [Boyer & Moore, 1979] describing THM, Boyer and Moore turned to the question of providing limited support for higher-order functions in their first-order setting. This had two very practical motivations. One was to allow the user to extend the prover by defining and mechanically verifying new proof procedures in the pure LISP dialect supported by THM. The other was to allow the user the convenience of LISP's "map functions" and LOOP facility. Both required formally defining the semantics of the logical language in the logic, i.e. axiomatizing the evaluation function EVAL. Ultimately this resulted in the provision of *metafunctions* [Boyer & Moore, 1981b] and the non-constructive "value-and-cost" function V&C$ [Boyer & Moore, 1988a], which were provided as part of the NQTHM system described in [Boyer & Moore, 1988b; 1998].

The most important side-effect of these additions, however, is under the hood; Boyer and Moore contrived to make the representation of constructor ground terms in the logic be identical to their representation as constants in its underlying implementation language LISP: integers are represented directly as LISP integers; for instance, s(s(s(0))) is represented by the machine-oriented internal LISP representation of 3, instead of the previous (ADD1 (ADD1 (ADD1 (ZERO)))). Symbols and list structures are embedded this way as well, so that they can can profit from the very efficient representation of these basic data types in LISP. It thus also became possible to represent symbolic machine states containing actual assembly code or the parse trees of actual programs in the logic of NQTHM. Metafunctions were put to good use canonicalizing symbolic state expressions. The exploration of formal operational semantics with NQTHM blossomed.

In addition, NQTHM adds a rational linear-arithmetic[175] decision procedure to the simplification stage of the waterfall [Boyer & Moore, 1988c], reducing the amount of user interaction necessary to prove arithmetic theorems. The incompleteness of the procedure when operating on terms beyond the linear fragment is of little practical importance since induction is available (and often automatic).

With NQTHM it became possible to formalize and verify problems beyond the scope of THM, such as the correctness of a netlist implementing the instruction-set architecture of a microprocessor [Hunt, 1985], Gödel's first incompleteness theorem,[176] the verified hard- and software stack of Computational Logic, Inc., relating a fabricated microprocessor design through an assembler, linker, loader, several compilers, and an operating system to simple verified application programs,[177] and the verification of the Berkeley C String Library.[178] Many more examples are listed in [Boyer & Moore, 1998].

---

[175]Linear arithmetic is traditionally called "Presburger Arithmetic" after Mojżesz Presburger (actually: "Prezburger") (1904–1943(?)); cf. [Presburger, 1930], [Stansifer, 1984], [Zygmunt, 1991].

6.5 ACL2

Because of the pervasive change in the representation of constants, the LISP subset supported by NQTHM is exponentially more efficient than the LISPs supported by THM and the PURE LISP THEOREM PROVER. It is still too inefficient, however: Emerging applications of NQTHM in the late 1980s included models of commercial microprocessors; users wished to run their models on industrial test suites. The root cause of the inefficiency was that ground execution in NQTHM was done by a purpose-built interpreter implemented by Boyer and Moore. To reach competitive speeds, it would have been necessary to build a good compiler and full run-time system for the LISP subset axiomatized in NQTHM. Instead, in August 1989, less than a year after the publication of [Boyer & Moore, 1988b] describing NQTHM, Boyer and Moore decided to axiomatize a practical subset of COMMON LISP [Steele, 1990], the then-emerging standard LISP, and to build an NQTHM-like theorem prover for it. To demonstrate that the subset was a practical programming language, they decided to code the theorem prover applicatively in that subset. Thus, ACL2 was born.

Boyer left Computational Logic, Inc., (CLI) and returned to his duties at the The University of Texas at Austin in 1989, while Moore resigned his tenure and stayed at CLI. This meant Moore was working full-time on ACL2, whereas Boyer was working on it only at night. Matt Kaufmann (*1952), who had worked with Boyer and Moore since the mid-1980s on NQTHM and had joined them at CLI, was invited to join the ACL2 project. By the mid-1990s, Boyer requested that his name be removed as an author of ACL2 because he no longer knew every line of code.

The only major change to inductive reasoning introduced by ACL2 is the further refinement of the induction templates computed at definition time: While NQTHM built the case analysis from the case conditions "governing" the recursive calls, ACL2 uses the more restrictive notion of the tests "ruling"[179] the recursive calls.

---

[176]Cf. [Shankar, 1994]. In [Shankar, 1994, p. xii] we read on this work with NQTHM:

"This theorem prover is known for its powerful heuristics for constructing proofs by induction while making clever use of previously proved lemmas. The Boyer–Moore theorem prover did not discover proofs of the incompleteness theorem but merely checked a detailed but fairly high-level proof containing over 2000 definitions and lemmas leading to the main theorems. These definitions and lemmas were constructed through a process of interaction with the theorem prover which was able to automatically prove a large number of nontrivial lemmas. By thus proving a well-chosen sequence of lemmas, the theorem prover is actually used as a *proof checker* rather than a theorem prover.

If we exclude the time spent thinking, planning, and writing about the proof, the verification of the incompleteness theorem occupied about eighteen months of effort with the theorem prover."

[177]Cf. [Moore, 1989b; 1989a], [Bevier &al., 1989], [Hunt, 1989], [Young, 1989], [Bevier, 1989].

ACL2 represents a major step, however, toward Boyer and Moore's dream of a *computational logic* because it is a theorem prover for a practical programming language. Because it is so used, *scaling* its algorithms and heuristics to deal with enormous models and the formulas they generate has been a major concern, as has been the efficiency of ground execution. Moreover, it also added many other proof techniques including congruence-based contextual rewriting, additional decision procedures, disjunctive search (meaning the waterfall no longer has just one pool but may generate several, one of which must be "emptied" to succeed), and many features made possible by the fact that the system code and state is visible to the logic and the user.

Among the landmark applications of ACL2 are the verification of a Motorola digital signal processor [Brock & Hunt, 1999] and of the floating-point division microcode for the AMD K5™ microprocessor [Moore &al., 1998], the routine verification of all elementary floating point arithmetic on the AMD Athlon™ [Russinoff, 1998], the certification of the Rockwell Collins AAMP7G™ for multi-level secure applications by the US National Security Agency based on the ACL2 proofs [Anon, 2005], and the integration of ACL2 into the work-flow of Centaur Technology, Inc., a major manufacturer of X86 microprocessors [Hunt & Swords, 2009]. Some of this work was done several years before the publications appeared because the early use of formal methods was considered proprietary.[180]

In most industrial applications of ACL2, induction is not used in every proof. Many of the proofs involve huge intermediate formulas, some requiring megabytes of storage simply to represent, let alone simplify. Almost all the proofs, however, depend on lemmas that require induction to prove.

To be successful, ACL2 must be good at both induction and simplification and *integrate* them seamlessly in a well-engineered system, so that the user can state and prove in a single system all the theorems needed.

ACL2 is most relevant to the historiography of inductive theorem proving because it demonstrates that the induction heuristics and the waterfall provide such an integration in ways that can be scaled to industrial-strength applications.

ACL2 and, by extension, inductive theorem proving, have changed the way microprocessors and low-level critical software are designed. Proof of correctness, or at least proof of some important system properties, is now a possibility.

---

[178]Via verification of its gcc-generated Motorola MC68020 machine code [Boyer & Yu, 1996].

[179]Compare the definition of *governors* on Page 180 of [Boyer & Moore, 1998] to the definition of *rulers* on Page 90 of [Kaufmann &al., 2000b].

[180]For example, the work for [Brock & Hunt, 1999] was completed in 1994, and that for [Moore &al., 1998] in 1995.

Boyer, Moore, and Kaufmann were awarded the 2005 ACM Software Systems Award for "the Boyer–Moore Theorem Prover":

> "The Boyer–Moore Theorem Prover is a highly engineered and effective formal-methods tool that pioneered the automation of proofs by induction, and now provides fully automatic or human-guided verification of critical computing systems. The latest version of the system, ACL2, is the only simulation/verification system that provides a standard modeling language and industrial-strength model simulation in a unified framework. This technology is truly remarkable in that simulation is comparable to C in performance, but runs inside a theorem prover that verifies properties by mathematical proof. ACL2 is used in industry by AMD, IBM, and Rockwell-Collins, among others."[181]

6.6 Further Historically Important Explicit-Induction Systems

Explicit induction is nowadays applied in many theorem proving systems, such as ISABELLE/HOL, COQ, PVS, and ISAPLANNER, to name just a few. We cannot treat all of these systems in this article. Thus, in this section, we sketch only those systems that provided crucial contributions to the history of the automation of mathematical induction.

6.6.1 RRL

RRL, the *Rewrite Rule Laboratory* [Kapur & Zhang, 1989], was initiated in 1982 and showed its main activity during its first dozen years. RRL is a system for proving the viability of many techniques related to term rewriting. Besides other forms of induction, RRL includes *cover-set induction*, which has eager induction-hypothesis generation, but is restricted to syntactic term orderings.

---

[181] For the complete text of the citation of Boyer, Moore, and Kaufmann see http://awards.acm.org/citation.cfm?id=4797627&aw=149.

6.6.2 INKA

The INKA project and the development of the INKA induction systems began at the University of Karlsruhe at the beginning of the 1980s. It became part of the Collaborative Research Center SFB 314 "Artificial Intelligence", which started in 1985 and was financed by the German Research Community (DFG) to overcome a backwardness in artificial intelligence in Germany of more than a decade compared to the research in Edinburgh and in the US.

The INKA systems were based on the concepts of Boyer & Moore [1979] and proved the executability of several new concepts, but they were never competitive with their contemporary Boyer–Moore theorem provers,[182] and the development of INKA was discontinued in the year 2000.

Three INKA system descriptions were presented at the CADE conference series: [Biundo &al., 1986], [Hutter & Sengler, 1996], [Autexier &al., 1999].

Besides interfaces to users and other systems, and the integration of logics, specifications, and results of other theorem provers, the essentially induction-relevant additions of INKA as compared to the system described in [Boyer & Moore, 1979] are the following: In [Biundo &al., 1986], there is an existential quantification where the system tries to find witnesses for the existentially quantified variables by interactive program synthesis. In [Hutter, 1994], there is synthesis of induction orderings by rippling (cf. § 7.2).

Most interesting work on explicit induction was realized along the line of the INKA systems: We have to mention here Christoph Walther's (*1950) elegant treatment of *automated termination proofs* for recursive function definitions [Walther, 1988; 1994b], and his theoretically outstanding work on the *generation of step cases* with eager induction-hypothesis generation [Walther, 1992; 1993]. Moreover, there is Dieter Hutter's (*1959) further development of *rippling* (cf. § 7.2), and Martin Protzen's (*1962) profound work on *patching of faulty conjectures* and on breaking out of the imagined cage of explicit induction by *"lazy induction"* [Protzen, 1994; 1995; 1996].

---

[182]INKA 5.0 [Autexier &al., 1999], however, was competitive in speed with NQTHM. This can roughly be concluded from the results of the inductive theorem proving contest at the 16th Int. Conf. on Automated Deduction (CADE), Trento (Italy), 1999 (the design of which is described in [Hutter & Bundy, 1999]), where the following systems competed with each other (in interaction with the following humans): NQTHM (Laurence Pierre), INKA 5.0 (Dieter Hutter), OYSTER/CLAM (Alan Bundy), and a first prototype of QUODLIBET (Ulrich Kühler). Only OYSTER/CLAM turned out to be significantly slower than the other systems, but all participating systems would have been left far behind ACL2 if it had participated.

6.6.3 OYSTER/CLAM

The OYSTER/CLAM system was developed at the University of Edinburgh in the late 1980s[183] and the 1990s by a large team led by Alan Bundy.[184]

OYSTER is a reimplementation of NUPRL [Constable &al., 1985], a proof editor for Martin-Löf constructive type theory with rules for structural induction in the style of Peano — a logic that is not well-suited for inductive proof search, as discussed in § 4.6. OYSTER is based on tactics with specifications in a meta-level language which provides a complete representation of the object level, but with a search space much better suited for inductive proof search.

CLAM is a proof planner (cf. § 7.1) which guides OYSTER, based on proof search in the meta-language, which includes rippling (cf. § 7.2).

OYSTER/CLAM is the slowest system explicitly mentioned in this article.[181] One reason for this inefficiency is its constructive object-level logic. Its successor systems, however, are much faster.[185]

In its line of development, OYSTER/CLAM proved the viability of several most important new concepts:

- Among the approaches that more or less address theorem proving in general, we have to mention *rippling* (cf. § 7.2) and a *productive use of failure* for the suggestion of crucial new lemmas.[186]

- A most interesting approach that addresses the core of the automation of inductive theorem proving and that deserves further development is the extension of recursion analysis to *ripple analysis*.[187]

---

[183] The system description [Bundy &al., 1990] of OYSTER/CLAM appeared already in summer 1990 at the CADE conference series (with a submission in winter 1989/1990); so the development must have started before the 1990s, contrary to what is stated in § 11.4 of [Bundy, 1999].

[184] For Alan Bundy see also Note 9.

[185] One of the much faster successor systems of OYSTER/CLAM under further development is ISAPLANNER, which is based on ISABELLE [Paulson, 1990]. See [Dixon & Fleuriot, 2003] and [Dennis &al., 2005] for early publications on ISAPLANNER.

[186] Cf. [Ireland & Bundy, 1994]. Moreover, see our discussion on the particular theoretical relevance of finding new lemmas in mathematical induction in § 4.10. Furthermore, note that the practical relevance of finding new lemmas addresses the efficiency of theorem proving in general, as described in Notes 72, 73, and 76 of § 4.10.

[187] Ripple analysis is sketched already in [Bundy &al., 1989, § 7] and nicely presented in [Bundy, 1999, § 7.10].

7 Alternative Approaches Besides Explicit Induction

In this section we will discuss the approaches to the automation of mathematical induction that do not strictly follow the method of explicit induction as we have described it. In general, these approaches are not disjoint from explicit induction. To the contrary, *proof planning* and *rippling* have until now been applied mostly to systems more or less based on explicit induction, but they are not exclusively related to induction and they are not following Boyer and Moore's method of explicit induction in every detail. Even systems for *implicit induction* may include many features of explicit induction and some of them actually do, such as RRL (cf. § 6.6.1) and QUODLIBET (cf. § 7.4).

7.1 Proof Planning

Suggestions on how to overcome an envisioned dead end in automated theorem proving were summarized in the end of the 1980s under the keyword *proof planning*. Besides its human-science aspects,[188] the main idea[189] of proof planning is to extend a theorem-proving system — on top of the *low-level search space* of the logic calculus of a proof checker — with a *higher-level search space*, which is typically smaller or better organized w.r.t. searching, more abstract, and more human-oriented.

The extensive and sophisticated subject of proof planning is not especially related to induction, but addresses automated theorem proving in general. We cannot cover it here and have to refer the reader to the standard publications on the subject.[190]

7.2 Rippling

Rippling is a technique for augmenting rewrite rules with information that helps to find a way to rewrite one expression (*goal*) into another (*target*), more specifically to reduce the difference between the goal and the target by rewriting the goal.

Although rippling is not restricted to inductive theorem proving, it was first used by Raymond Aubin[191] in the context of the description of heuristics for the automation of mathematical induction and found most of its applications there.

---

[188] Cf. [Bundy, 1989].

[189] Cf. [Bundy, 1988], [Dennis &al., 2005].

[190] In addition to [Bundy, 1988; 1989] and [Dennis &al., 2005], see also [Dietrich, 2011], [Melis &al., 2008], [Jamnik &al., 2003], and the references there.

[191] The verb "to ripple up" is used in §§ 3.2 and 3.4 of [Aubin, 1976] — not as a technical term, but just as an informal term for motivating some heuristics. The formalizers of rippling give explicit credit to Aubin [1976] for their inspiration in [Bundy &al., 2005, § 1.10, p. 21], although Aubin does not mention the term at any other place in his publications [Aubin, 1976; 1979]. Note, however, that instead of today's name "rippling out", Aubin actually used "rippling up".

The leading developers and formalizers of the technique are Alan Bundy, Dieter Hutter, David Basin, Frank van Harmelen, and Andrew Ireland.

We have already mentioned rippling in § 6.6 several times, but this huge and well-documented area of research cannot be covered here, and we have to refer the reader to the monograph [Bundy &al., 2005].[192]

Let us explain here, however, why rippling can be most helpful in the automation of simple inductive proofs.

Roughly speaking, the remarkable success in proving *simple* theorems by induction automatically, can be explained as follows: If we look upon the task of proving a theorem as reducing it to a tautology, then we have more heuristic guidance when we know that we probably have to do it by mathematical induction: Tautologies can have arbitrary subformulas, but the induction hypothesis we are going to apply can restrict the search space tremendously.

In a cartoon of Alan Bundy's, the original theorem is pictured as a zigzagged mountainscape and the reduced theorem after the unfolding of recursive operators according to recursion analysis (goal) is pictured as the reflection of the mountainscape on the surface of a lake with ripples. To apply the induction hypothesis (target), instead of the uninformed search for an arbitrary tautology, we have to *get rid of the ripples* to be able to apply an instance of the theorem as induction hypothesis to the mountainscape mirrored by the calmed surface of the lake.

A crucial advantage of rippling in the area of automated induction is that it can also be used to suggest missing lemmas as described in [Ireland & Bundy, 1994].

Until today, rippling was applied to the automation of induction only within explicit induction, whereas it is clearly not limited to explicit induction, and we actually expect it to be more useful in areas of automated theorem proving with bigger search spaces and, in particular, in *descente infinie*.

7.3 Implicit Induction

The remaining approaches to mechanize mathematical induction *not* subsumed by explicit induction, however, are united under the name "implicit induction".

Triggered[193] by the success of Boyer & Moore [1979], publication on these alternative approaches started already in the year 1980 in purely equational theories.[194] A sequence of papers on technical improvements[195] was topped by [Bachmair, 1988], which gave rise to a hope to develop the method into practical usefulness, although it was still restricted to purely equational theories.

---

[192]Historically important are also the following publications on rippling: [Hutter, 1990], [Bundy &al., 1991], [Ireland & Bundy, 1994], [Basin & Walsh, 1996].

Inspired by [Bachmair, 1988], in the late 1980s and the first half of the 1990s several researchers tried to understand more clearly what implicit induction means from a theoretical point of view and whether it could be useful in practice.[196]

While it is generally accepted that [Bachmair, 1988] is about implicit induction and [Boyer & Moore, 1979] is about explicit induction, there are the following three different viewpoints on what the essential aspect of implicit induction actually is.

Proof by Consistency:[197] Systems for proof by consistency run some Knuth–Bendix[198] or superposition[199] completion procedure.

A proof attempt is successful when the prover has drawn all necessary inferences and stops without having detected any inconsistency.

The runs are typically infinite, however, and the admissibility conditions are too restrictive for most applications.

Even in the rare case that it stops, proof by consistency has shown to perform far worse than any other known form of mechanizing mathematical induction, mainly because it requires the generation of far too many irrelevant and superfluous inferences, under which the ones relevant for establishing the induction steps can hardly be made *explicit*.

Roughly speaking, the conceptual flaw of proof by consistency is that, instead of finding a sufficient set of reasonable inferences, the research follows the idea of ruling out as many irrelevant inferences as possible.

---

[193] Although it is obvious that in the relatively small community of artificial intelligence and computer science in the 1970s, the success of [Boyer & Moore, 1979] triggered the publication of papers on induction in the term rewriting community, we can document the influence of Boyer and Moore's work here only with the following facts: [Boyer & Moore, 1975; 1979] are both cited in [Huet & Hullot, 1980]. [Boyer & Moore, 1977] is cited in [Musser, 1980] as one of the "important sources of inspiration". Moreover, Lankford [1980] constitutively refers to a personal communication with Robert S. Boyer in 1979. Finally, Goguen [1980] avoids a direct reference to Boyer and Moore, but cites only the PhD thesis [Aubin, 1976] of Raymond Aubin, following their work in Edinburgh.

[194] Cf. [Goguen, 1980], [Huet & Hullot, 1980], [Lankford, 1980], [Musser, 1980].

[195] Cf. [Göbel, 1985], [Jouannaud & Kounalis, 1986], [Fribourg, 1986], [Küchlin, 1989].

[196] Cf. e.g. [Zhang &al., 1988], [Kapur & Zhang, 1989], [Bevers & Lewi, 1990], [Reddy, 1990], [Gramlich & Lindner, 1991], [Ganzinger & Stuber, 1992], [Bouhoula & Rusinowitch, 1995], [Padawitz, 1996].

[197] The name "proof by consistency" was coined in the title of [Kapur & Musser, 1987], which is the later published forerunner of its outstanding improved version [Kapur & Musser, 1986].

[198] See UNICOM [Gramlich & Lindner, 1991] for such a system, following [Bachmair, 1988] with several improvements. See [Knuth & Bendix, 1970] for the Knuth–Bendix completion procedure.

[199] See [Ganzinger & Stuber, 1992] for such a system.

Implicit Induction Ordering: In the early implicit-induction systems,[200] induction proceeds over a syntactic term ordering, which typically cannot be made *explicit* in the sense that there would be some predicate term in the logical syntax that denotes this ordering in the intended models of the specification. The semantic orderings of explicit induction, however, cannot depend on the precise syntactic term structure of a weight w, but only on the value of w under an evaluation in the intended models.

In contrast to rudimentary inference systems that turned out to be more or less useless in practice (such as the one of [Bachmair, 1988] for inductive completion in unconditional equational specifications), more powerful human-oriented inference systems (such as the one of QUODLIBET) are considerably restrained by the constraint to be sound also for induction orderings that depend on the precise syntactic structure of terms (beyond their values).[201]

The early implicit-induction systems needed such sophisticated term orderings,[202] because they started from the induction conclusion and every inference step reduced the formulas w.r.t. the induction ordering again and again, but an application of an induction hypothesis was admissible to greater formulas only. This deterioration of the ordering information with every inference step was overcome by the introduction of explicit weight terms in [Wirth & Becker, 1995], which obviate the former need for syntactic term orderings as induction orderings.

Descente Infinie ("Lazy Induction"): Contrary to explicit induction, where induction is introduced into an otherwise merely deductive inference system only by the *explicit* application of induction axioms in the induction rule, the cyclic arguments and their well-foundedness in implicit induction need not be confined to single inference steps.[203] The induction rule of explicit induction generates all induction hypotheses in a single inference step. To the contrary, in implicit induction, the inference system "knows" what an induction hypothesis is, i.e. it includes inference rules that provide or apply induction hypotheses, given that certain ordering conditions resulting from these applications can be met by an induction ordering. Because this aspect of implicit induction can facilitate the human-oriented induction method described in § 4.6,

---

[200] See [Gramlich & Lindner, 1991] and [Ganzinger & Stuber, 1992] for such systems.

[201] This soundness constraint (still observed in [Wirth, 1997]) was dropped in the further development of QUODLIBET in [Kühler, 2000], because it turned out to be unintuitive and superfluous.

[202] Cf. e.g. [Bachmair, 1988], [Steinbach, 1988; 1995], [Geser, 1996].

[203] For this reason, the funny name "inductionless induction" was originally coined for implicit induction in the titles of [Lankford, 1980; 1981] as a short form for "induction without induction rule". See also the title of [Goguen, 1980] for a similar phrase.

the name *descente infinie* was coined for it (cf. § 4.7). Researchers introduced to this aspect by [Protzen, 1994] (entitled "Lazy Generation of Induction Hypotheses") sometimes speak of "lazy induction" instead of *descente infinie*.

The entire handbook article [Comon, 2001] (with corrections in [Wirth, 2005a]) is dedicated to the two aspects of *proof by consistency* and *implicit induction orderings*. Today, however, the interest in these two aspects tends to be historical or theoretical, especially because these aspects can hardly be combined with explicit induction.

In contrast, *descente infinie* synergetically combines with explicit induction, as witnessed by the QUODLIBET system, which we will discuss in § 7.4.

7.4 QUODLIBET

In the last years of the Collaborative Research Center SFB 314 "Artificial Intelligence" (cf. § 6.6.2), after extensive experiments with several inductive theorem proving systems,[204] such as the explicit-induction systems NQTHM (cf. § 6.4) and INKA (cf. § 6.6.2), the implicit-induction system UNICOM [Gramlich & Lindner, 1991], and the mixed system RRL (cf. § 6.6.1), Claus-Peter Wirth (*1963) and Ulrich Kühler (*1964) came to the conclusion that — in spite of the excellent interaction concept of UNICOM[205] — *descente infinie* was actually the only aspect of implicit induction that deserved further investigation. Moreover, the coding of recursive functions in *unconditional* equations in UNICOM turned out to be most inadequate for inductive theorem proving in practice, where positive/negative-conditional equations were in demand for specification, as well as clausal logic for theorem proving.[206]

Therefore, a new system had to be created, which was given the name QUODLIBET (Latin for "as you like it"), because it should enable its users to avoid overspecification by admitting partial function specifications, and to execute proofs whose crucial proof steps mirror exactly the intended ones.[207]

A concept for partial function specification instead of the totality requirement of explicit induction was easily obtained by elaborating the first part of [Wirth, 1991] into the framework for positive/negative-conditional rewrite systems of [Wirth & Gramlich, 1994a]. After inventing constructor variables in [Wirth &al., 1993], the monotonicity of validity w.r.t. consistent extension of the partial specifications was easily achieved [Wirth & Gramlich, 1994b], so that the induction proofs did not have to be re-done after such an extension of a partially defined function.

---

[204] Cf. [Kühler, 1991].
[205] For the assessment of UNICOM's interaction concept see [Kühler, 1991, p. 134ff.].
[206] See [Kühler, 1991, pp. 134, 138].

Although the efficiently decidable confluence criterion that defines admissibility of function definitions in QUODLIBET and guarantees their (object-level) consistency (cf. § 5.2) was very hard to prove and was presented completely and in an appropriate form not before [Wirth, 2009], the essential admissibility requirements were already clear in 1996.[208]

The weak admissibility conditions of QUODLIBET — mutually recursive functions, possibly partially defined because of missing cases or non-termination — are of practical importance. Although humans can code mutually recursive functions into non-mutually recursive functions,[209] they will hardly be able to understand complicated formulas where these encodings occur, and so they will have severe problems in assisting the proving system in the construction of hard proofs. Partiality due to non-termination essentially occurs in interpreters with undecidable domains. Partiality due to missing cases of the definition can often be avoided by overspecification in theory, but not in practice where the unintended results of overspecification may complicate matters considerably.

For instance, Bernd Löchner (*1967) (a user, not a developer of QUODLIBET) concludes in [Löchner, 2006, p. 76]:

> "The translation of the different specifications into the input language of the inductive theorem prover QUODLIBET [Avenhaus &al., 2003] was straightforward. We later realized that this is difficult or impossible with several other inductive provers as these have problems with mutual recursive functions and partiality" ...

---

[207]We cannot claim that QUODLIBET is actually able to execute proofs whose crucial proof steps mirror exactly the ones intended by its human users, simply because this was not scientifically investigated, say in terms of cognitive psychology. Users, however, considered it to be more appropriate than other systems in this aspect, mostly due to the direct support for partial and mutually recursive function specification, cf. [Löchner, 2006]. Moreover, the four dozen elementary rules of QUODLIBET's inference machine were designed to mirror the way human's organize their proofs (cf. [Wirth, 1997], [Kühler, 2000]); so a user has to deal with one natural inference step where OYSTER may have hundreds of intuitionistic steps. The appropriateness of QUODLIBET's calculus for interchanging information with humans deteriorated, however, after adding inference rules for the efficient implementation of Presburger Arithmetic. Note that the calculus is only the lowest logic level a user of a theorem-proving system may have to deal with; from our experience with many such systems we came to the firm conviction, however, that the automation of proof search will always fail on the lowest logic level from time to time, such that human-oriented state-of-the-art logic calculi are essential for the acceptance of automated, interactive theorem provers by their users.

[208]See [Kühler & Wirth, 1996] for the first publication of the object-level consistency of the specifications that are admissible and supported with strong induction heuristics in QUODLIBET. In [Kühler & Wirth, 1996], a huge proof from the original 1995 edition of [Wirth, 2005b] guaranteed the consistency. Moreover, the most relevant and appropriate one of the seven inductive validities of [Wirth & Gramlich, 1994b] is chosen for QUODLIBET in [Kühler & Wirth, 1996] (no longer the initial or free models typical for implicit induction!).

Based on the *descente infinie* inference system for clausal first-order logic of [Wirth & Kühler, 1995],[210] the system development of QUODLIBET in COMMON LISP (cf. § 6.5), mostly by Kühler and Tobias Schmidt-Samoa (*1973), lasted from 1995 to 2006. The system was described and demonstrated at the 19th Int. Conf. on Automated Deduction (CADE), Miami Beach (FL), 2003 [Avenhaus &al., 2003]. The extension of the *descente infinie* inference systems of QUODLIBET to the full [modal] higher-order logic of [Wirth, 2004; 2017] has not been implemented yet.

To the best of our knowledge, QUODLIBET is the first theorem prover whose proof state is an and-or-tree (of clauses); actually, a forest of such trees, so that each conjecture that can generate induction hypotheses in a possibly mutual induction proof has its own tree [Kühler, 2000]. An extension of the recursion analysis of [Boyer & Moore, 1979] for constructor-style specifications (cf. § 5.5) was developed by writing and testing tactics in QUODLIBET's PASCAL-like[211] meta-language QML [Kühler, 2000]. To achieve an acceptable run-time performance (but not competitive with ACL2, of course), QML tactics are compiled before execution.

In principle, termination proofs are not required, simply because termination is not an admissibility requirement in QUODLIBET. Instead, definition-time recursion analysis uses induction lemmas (cf. § 6.3.7) to prove lemmas on function domains by induction.[212]

At proof time, recursion analysis is used by the standard tactic only to determine the induction variables from the induction templates: As seen in Example 4.3 of § 4.7 w.r.t. the strengthened transitivity of lessp (as compared to the explicit-induction proof in Example 6.2 of § 6.2.6 and Example 6.13 of § 6.3.8), subsumption and merging of schemes are not required in *descente infinie*.[213]

A considerable speed-up of QUODLIBET and an extension of its automatically provable theorems was achieved by Schmidt-Samoa during his PhD work with the system in 2004–2006. He developed a marking concept for the tagging of rewrite lemmas (cf. § 6.3.1), where the elements of a clause can be marked as Forbidden,

---

[209]See the first paragraph of § 5.7.

[210]Later improvements of this inference system are found in [Wirth, 1997], [Kühler, 2000], and [Schmidt-Samoa, 2006b].

[211]See [Wirth, 1971] for the programming language PASCAL. The critical decision for an imperative instead of a functional tactics language turned out to be most appropriate during the ten years of using QML.

[212]While domain lemmas for totally defined functions are usually found without interaction and total functions do not provide relevant overhead in QUODLIBET, the user often has to help in case of partial function definitions by providing *domain lemmas* such as
$$\text{Def delfirst}(x,l), \quad \text{mbp}(x,l) \neq \text{true},$$
for delfirst defined via (delfirst1–2) of § 4.5.

Mandatory, Obligatory, and Generous, to control the recursive relief of conditions in contextual rewriting [Schmidt-Samoa, 2006b; 2006c]. Moreover, a very simple, but most effective reuse mechanism analyzes during a proof attempt whether it actually establishes a proof of some sub-clause, and uses this knowledge to crop conjunctive branches that do not contribute to the actual goal [Schmidt-Samoa, 2006b]. Finally, an even closer integration of linear arithmetic (cf. Note 175) with excellent results [Schmidt-Samoa, 2006a; 2006b] questioned one of the basic principles of QUODLIBET, namely the idea that the prover does not try to be clever, but stops early if there is no progress visible, and presents the human user the proof state in a nice graphical tree representation: The expanded highly-optimized formulation of arithmetic by means of special functions for the decidable fragment of Presburger Arithmetic results in clauses that do not easily admit human inspection anymore. We did not find means to overcome this, because we did not find a way to fold theses clauses to achieve a human-oriented higher level of abstraction.

QUODLIBET is, of course, able to do all[214] *descente infinie* proofs of our examples automatically. Moreover, QUODLIBET finds all proofs for the irrationality of the square root of two indicated in Figure 3 (sketched in § 6.3.9) automatically and without explicit hints on the induction ordering (say, via newly defined nonsensical functions, such as the one given in (sqrtio1) of § 6.3.9) — provided that the required lemmas are available.

All in all, QUODLIBET has proved that *descente infinie* ("lazy induction") goes well together with explicit induction and that we have reason to hope that eager induction-hypotheses generation can be overcome for theorems with difficult induction proofs, sacrificing neither efficiency nor the usefulness of the excellent heuristic knowledge developed in explicit induction. Why *descente infinie* and human-orientedness should remain on the agenda for induction in mathematics assistance systems is explained in the manifesto [Wirth, 2012c].

---

[213]Although it is not a must and not part of the standard tactic, induction hypotheses may be generated eagerly in QUODLIBET to enhance generalization as in Example 4.5 of § 4.9, in which case subsumption and merging of induction schemes as described in § 6.3.8 are required. Moreover, the concept of flawed induction schemes of QUODLIBET (taken over from THM as well, cf. § 6.3.8) depends on the mergeability of schemes. Furthermore, QUODLIBET actually applies some merging techniques to plan case analyses optimized for induction [Kühler, 2000, § 8.3.3]. The question why QUODLIBET adopts the great ideas of recursion analysis from THM, but does not follow them precisely, has two answers: First, it was necessary to extend the heuristics of THM to deal with constructor-style definitions. The second answer was already given in § 6.3.9: Testing is the only judge on heuristics.

[214]These three *descente infinie* proofs are presented as Examples 4.2 and 4.3 of § 4.7, and Example 4.5 of § 4.9.

8 Conclusion

"One of the reasons our theorem prover is successful is that we trick the user into telling us the proof. And the best example of that, that I know, is: If you want to prove that there exists a prime factorization — that is to say a list of primes whose product is any given number — then the way you state it is: You define a function that takes a natural number and delivers a list of primes, and then you prove that it does that. And, of course, the definition of that function is everybody else's proof. The absence of quantifiers and the focus on constructive, you know, recursive definitions forces people to do the work. And so then, when the theorem prover proves it, they say 'Oh what wonderful theorem prover!', without even realizing they sweated bullets to express the theorem in that impoverished logic."

said Moore, and Boyer agreed laughingly.[215]

Acknowledgments

We would like to thank Fabio Acerbi, Klaus Barner, Anne O. Boyer, Robert S. Boyer, Alan Bundy, Catherine Goldstein, Bernhard Gramlich, Warren A. Hunt, Dieter Hutter, Matt Kaufmann, Ulrich Kühler, Klaus Madlener, Jo Moore, Peter Padawitz, Tobias Schmidt-Samoa, Jörg Siekmann, Judith Stengel, and Marianne Wirth.

In particular, we are obliged to our second reader Alan Bundy for his most careful reading and long long list of flaws and elaborate suggestions that improved this article considerably.

As our our second reader Bernhard Gramlich (1959–2014), a gifted teacher, a most creative and scrutinous researcher, and a true friend, passed away — much too early — before dawn on June 3, 2014, we would like to dedicate this article to him. Besides the automation of mathematical induction, Bernhard's main focus was on confluence and termination of term rewriting systems, where he proved many of the hardest theorems of the area [Gramlich, 1996].

---

[215][Wirth, 2012d].

References

[Abrahams &al., 1980] Paul W. Abrahams, Richard J. Lipton, and Stephen R. Bourne, editors, 1980. *Conference Record of the 7th Annual ACM SIGPLAN-SIGACT Symposium on Principles of Programming Languages (POPL), Las Vegas (NV), 1980.* ACM Press. http://dl.acm.org/citation.cfm?id=567446.

[Acerbi, 2000] Fabio Acerbi, 2000. Plato: Parmenides 149a7–c3. A proof by complete induction? *Archive for History of Exact Sciences*, 55:57–76.

[Ackermann, 1928] Wilhelm Ackermann, 1928. Zum Hilbertschen Aufbau der reellen Zahlen. *Mathematische Annalen*, 99:118–133. Received Jan. 20, 1927.

[Ackermann, 1940] Wilhelm Ackermann, 1940. Zur Widerspruchsfreiheit der Zahlentheorie. *Mathematische Annalen*, 117:163–194. Received Aug. 15, 1939.

[Aït-Kaci & Nivat, 1989] Hassan Aït-Kaci and Maurice Nivat, editors, 1989. *Proc. of the Colloquium on Resolution of Equations in Algebraic Structures (CREAS), Lakeway (TX), 1987.* Academic Press (Elsevier).

[Anon, 1899] Anon, editor, 1899. *Festschrift zur Feier der Enthüllung des Gauß-Weber-Denkmals in Göttingen, herausgegeben von dem Fest-Comitee.* Verlag von B. G. Teubner, Leipzig.

[Anon, 2005] Anon, 2005. Advanced Architecture MicroProcessor 7 Government (AAMP7G) microprocessor. Rockwell Collins, Inc. WWW only: http://www.rockwellcollins.com/sitecore/content/Data/Products/Information_Assurance/Cryptography/AAMP7G_Microprocessor.aspx.

[Armando &al., 2008] Alessandro Armando, Peter Baumgartner, and Gilles Dowek, editors, 2008. *4th Int. Joint Conf. on Automated Reasoning (IJCAR), Sydney (Australia), 2008*, number 5195 in Lecture Notes in Artificial Intelligence. Springer.

[Aubin, 1976] Raymond Aubin, 1976. *Mechanizing Structural Induction.* PhD thesis, Univ. Edinburgh. Short version is [Aubin, 1979]. http://hdl.handle.net/1842/6649.

[Aubin, 1979] Raymond Aubin, 1979. Mechanizing Structural Induction — Part I: Formal System. Part II: Strategies. *Theoretical Computer Sci.*, 9:329–345+347–362. Received March (Part I) and November (Part II) 1977, rev. March 1978. Long version is [Aubin, 1976].

[Autexier, 2005] Serge Autexier, 2005. On the dynamic increase of multiplicities in matrix proof methods for classical higher-order logic. In [Beckert, 2005, pp. 48–62].

[Autexier &al., 1999] Serge Autexier, Dieter Hutter, Heiko Mantel, and Axel Schairer, 1999. System description: INKA 5.0 – a logical voyager. In [Ganzinger, 1999, pp. 207–211].

[Avenhaus &al., 2003] Jürgen Avenhaus, Ulrich Kühler, Tobias Schmidt-Samoa, and Claus-Peter Wirth, 2003. How to prove inductive theorems? QUODLIBET! In [Baader, 2003, pp. 328–333], http://wirth.bplaced.net/p/quodlibet.

[Baader, 2003] Franz Baader, editor, 2003. *19th Int. Conf. on Automated Deduction (CADE), Miami Beach (FL), 2003*, number 2741 in Lecture Notes in Artificial Intelligence. Springer.

[Baaz & Leitsch, 1995] Matthias Baaz and Alexander Leitsch, 1995. Methods of functional extension. *Collegium Logicum — Annals of the Kurt Gödel Society*, 1:87–122.

[Bachmair, 1988] Leo Bachmair, 1988. Proof by consistency in equational theories. In [LICS, 1988, pp. 228–233].

[Bachmair &al., 1992] Leo Bachmair, Harald Ganzinger, and Wolfgang J. Paul, editors,

1992. *Informatik – Festschrift zum 60. Geburtstag von Günter Hotz*. B. G. Teubner Verlagsgesellschaft.

[Bajscy, 1993] Ruzena Bajscy, editor, 1993. *Proc. 13th Int. Joint Conf. on Artificial Intelligence (IJCAI), Chambery (France)*. Morgan Kaufmann (Elsevier), Los Altos (CA). http://ijcai.org/Past%20Proceedings.

[Barendregt, 1981] Hen(dri)k P. Barendregt, 1981. *The Lambda Calculus — Its Syntax and Semantics*. Number 103 in Studies in Logic and the Foundations of Mathematics. North-Holland (Elsevier). 1st edn. (final rev. edn. is [Barendregt, 2012]).

[Barendregt, 2012] Hen(dri)k P. Barendregt, 2012. *The Lambda Calculus — Its Syntax and Semantics*. Number 40 in Studies in Logic. College Publications, London. 6th rev. edn. (1st edn. is [Barendregt, 1981]).

[Barner, 2001a] Klaus Barner, 2001a. Pierre Fermat (1601?–1665) — His life beside mathematics. *European Mathematical Society Newsletter*, 43 (Dec. 2001):12–16. Long version in German is [Barner, 2001b]. www.ems-ph.org/journals/newsletter/pdf/2001-12-42.pdf.

[Barner, 2001b] Klaus Barner, 2001b. Das Leben Fermats. *DMV-Mitteilungen*, 3/2001:12–26. Extensions in [Barner, 2007]. Short versions in English are [Barner, 2001c; 2001a].

[Barner, 2001c] Klaus Barner, 2001c. How old did Fermat become? *NTM Internationale Zeitschrift für Geschichte und Ethik der Naturwissenschaften, Technik und Medizin, Neue Serie, ISSN 00366978*, 9:209–228. Long version in German is [Barner, 2001b]. New results on the subject in [Barner, 2007].

[Barner, 2007] Klaus Barner, 2007. Neues zu Fermats Geburtsdatum. *DMV-Mitteilungen*, 15:12–14. *(Further support for the results of [Barner, 2001c], narrowing down Fermat's birth date from 1607/8 to Nov. 1607)*.

[Basin & Walsh, 1996] David Basin and Toby Walsh, 1996. A calculus for and termination of rippling. *J. Automated Reasoning*, 16:147–180.

[Becker, 1965] Oscar Becker, editor, 1965. *Zur Geschichte der griechischen Mathematik*. Wissenschaftliche Buchgesellschaft, Darmstadt.

[Beckert, 2005] Bernhard Beckert, editor, 2005. *14th Int. Conf. on Tableaux and Related Methods, Koblenz (Germany), 2005*, number 3702 in Lecture Notes in Artificial Intelligence. Springer.

[Bell & Thayer, 1976] Thomas E. Bell and T. A. Thayer, 1976. Software requirements: Are they really a problem? In [Yeh & Ramamoorthy, 1976, pp. 61–68], http://pdf.aminer.org/000/361/405/software_requirements_are_they_really_a_problem.pdf.

[Benzmüller &al., 2008] Christoph Benzmüller, Frank Theiss, Lawrence C. Paulson, and Arnaud Fietzke, 2008. Leo-II — a cooperative automatic theorem prover for higher-order logic. In [Armando &al., 2008, pp. 162–170].

[Berka & Kreiser, 1973] Karel Berka and Lothar Kreiser, editors, 1973. *Logik-Texte – Kommentierte Auswahl zur Geschichte der modernen Logik*. Akademie Verlag GmbH, Berlin. 2nd rev. edn. (1st edn. 1971; 4th rev. rev. edn. 1986).

[Bernays, 1928] Paul Bernays, 1928. Zusatz zu Hilberts Vortrag „Die Grundlagen der Mathematik". *Abhandlungen aus dem mathematischen Seminar der Univ. Hamburg*, 6:89–92. English translation *On the Consistency of Arithmetic* in [Heijenoort, 1971, pp. 485–489].

[Bevers & Lewi, 1990] Eddy Bevers and Johan Lewi, 1990. Proof by consistency in conditional equational theories. Tech. Report CW 102, Dept. Comp. Sci., K. U. Leuven. Rev. July 1990. Short version in [Kaplan & Okada, 1991, pp. 194–205].

[Bevier, 1989] William R. Bevier, 1989. Kit and the short stack. *J. Automated Reasoning*, 5:519–530.

[Bevier &al., 1989] William R. Bevier, Warren A. Hunt, J Strother Moore, and William D. Young, 1989. An approach to systems verification. *J. Automated Reasoning*, 5:411–428.

[Bibel & Kowalski, 1980] Wolfgang Bibel and Robert A. Kowalski, editors, 1980. *5th Int. Conf. on Automated Deduction (CADE), Les Arcs (France), 1980*, number 87 in Lecture Notes in Computer Science. Springer.

[Biundo &al., 1986] Susanne Biundo, Birgit Hummel, Dieter Hutter, and Christoph Walther, 1986. The Karlsruhe inductive theorem proving system. In [Siekmann, 1986, pp. 673–675].

[Bledsoe, 1971] W. W. Bledsoe, 1971. Splitting and reduction heuristics in automatic theorem proving. *Artificial Intelligence*, 2:55–77.

[Bledsoe &al., 1971] W. W. Bledsoe, Robert S. Boyer, and William H. Henneman, 1971. Computer proofs of limit theorems. In [Cooper, 1971, pp. 586–600]. Long version is [Bledsoe &al., 1972].

[Bledsoe &al., 1972] W. W. Bledsoe, Robert S. Boyer, and William H. Henneman, 1972. Computer proofs of limit theorems. *Artificial Intelligence*, 3:27–60. Short version is [Bledsoe &al., 1971].

[Bledsoe & Loveland, 1984] W. W. Bledsoe and Donald W. Loveland, editors, 1984. *Automated Theorem Proving: After 25 Years*. Number 29 in Contemporary Mathematics. American Math. Soc., Providence (RI). Proc. of the Special Session on Automatic Theorem Proving, 89th Annual Meeting of the American Math. Soc., Denver (CO), Jan. 1983.

[Bouajjani & Maler, 2009] Ahmed Bouajjani and Oded Maler, editors, 2009. *Proc. 21st Int. Conf. on Computer Aided Verification (CAV), Grenoble (France), 2009*, volume 5643 of *Lecture Notes in Computer Science*. Springer.

[Bouhoula & Rusinowitch, 1995] Adel Bouhoula and Michaël Rusinowitch, 1995. Implicit induction in conditional theories. *J. Automated Reasoning*, 14:189–235.

[Bourbaki, 1939] Nicolas Bourbaki, 1939. *Éléments des Mathématique — Livre 1: Théorie des Ensembles. Fascicule De Résultats*. Number 846 in Actualités Scientifiques et Industrielles. Hermann, Paris. 1st edn., VIII + 50 pp.. Review is [Church, 1946]. 2nd rev. extd. edn. is [Bourbaki, 1951].

[Bourbaki, 1951] Nicolas Bourbaki, 1951. *Éléments des Mathématique — Livre 1: Théorie des Ensembles. Fascicule De Résultats*. Number 846-1141 in Actualités Scientifiques et Industrielles. Hermann, Paris. 2nd rev. extd. edn. of [Bourbaki, 1939]. 3rd rev. extd. edn. is [Bourbaki, 1958b].

[Bourbaki, 1954] Nicolas Bourbaki, 1954. *Éléments des Mathématique — Livre 1: Théorie des Ensembles. Chapitre I & II*. Number 1212 in Actualités Scientifiques et Industrielles. Hermann, Paris. 1st edn.. 2nd rev. edn. is [Bourbaki, 1960].

[Bourbaki, 1956] Nicolas Bourbaki, 1956. *Éléments des Mathématique — Livre 1: Théorie des Ensembles. Chapitre III*. Number 1243 in Actualités Scientifiques et Industrielles. Hermann, Paris. 1st edn., II + 119 + 4 (mode d'emploi) + 23 (errata no. 6) pp.. 2nd rev. extd. edn. is [Bourbaki, 1967].

[Bourbaki, 1958a] Nicolas Bourbaki, 1958a. *Éléments des Mathématique — Livre 1: Théorie des Ensembles. Chapitre IV*. Number 1258 in Actualités Scientifiques et Industrielles. Hermann, Paris. 1st edn.. 2nd rev. extd. edn. is [Bourbaki, 1966a].

[Bourbaki, 1958b] Nicolas Bourbaki, 1958b. *Éléments des Mathématique — Livre 1: Théorie des Ensembles. Fascicule De Résultats*. Number 1141 in Actualités Scientifiques et Industrielles. Hermann, Paris. 3rd rev. extd. edn. of [Bourbaki, 1951]. 4th rev. extd. edn. is [Bourbaki, 1964].

[Bourbaki, 1960] Nicolas Bourbaki, 1960. *Éléments des Mathématique — Livre 1: Théorie des Ensembles. Chapitre I & II*. Number 1212 in Actualités Scientifiques et Industrielles. Hermann, Paris. 2nd rev. extd. edn. of [Bourbaki, 1954]. 3rd rev. edn. is [Bourbaki, 1966b].

[Bourbaki, 1964] Nicolas Bourbaki, 1964. *Éléments des Mathématique — Livre 1: Théorie des Ensembles. Fascicule De Résultats*. Number 1141 in Actualités Scientifiques et Industrielles. Hermann, Paris. 4th rev. extd. edn. of [Bourbaki, 1958b]. 5th rev. extd. edn. is [Bourbaki, 1968b].

[Bourbaki, 1966a] Nicolas Bourbaki, 1966a. *Éléments des Mathématique — Livre 1: Théorie des Ensembles. Chapitre IV*. Number 1258 in Actualités Scientifiques et Industrielles. Hermann, Paris. 2nd rev. extd. edn. of [Bourbaki, 1958a]. English translation in [Bourbaki, 1968a].

[Bourbaki, 1966b] Nicolas Bourbaki, 1966b. *Éléments des Mathématique — Livre 1: Théorie des Ensembles. Chapitres I & II*. Number 1212 in Actualités Scientifiques et Industrielles. Hermann, Paris. 3rd rev. edn. of [Bourbaki, 1960], VI + 143 + 7 (errata no. 13) pp.. English translation in [Bourbaki, 1968a].

[Bourbaki, 1967] Nicolas Bourbaki, 1967. *Éléments des Mathématique — Livre 1: Théorie des Ensembles. Chapitre III*. Number 1243 in Actualités Scientifiques et Industrielles. Hermann, Paris. 2nd rev. extd. edn. of [Bourbaki, 1956], 151 + 7 (errata no. 13) pp.. 3rd rev. edn. results from executing these errata. English translation in [Bourbaki, 1968a].

[Bourbaki, 1968a] Nicolas Bourbaki, 1968a. *Elements of Mathematics — Theory of Sets*. Actualités Scientifiques et Industrielles. Hermann, Paris, jointly published with AdiWes International Series in Mathematics, Addison–Wesley, Reading (MA). English translation of [Bourbaki, 1966b; 1967; 1966a; 1968b].

[Bourbaki, 1968b] Nicolas Bourbaki, 1968b. *Éléments des Mathématique — Livre 1: Théorie des Ensembles. Fascicule De Résultats*. Number 1141 in Actualités Scientifiques et Industrielles. Hermann, Paris. 5th rev. extd. edn. of [Bourbaki, 1964]. English translation in [Bourbaki, 1968a].

[Boyer, 1971] Robert S. Boyer, 1971. *Locking: a restriction of resolution*. PhD thesis, The University of Texas at Austin.

[Boyer, 2012] Robert S. Boyer, 2012. E-mail to Claus-Peter Wirth, Nov. 19,

[Boyer &al., 1973] Robert S. Boyer, D. Julian M. Davies, and J Strother Moore, 1973. The 77-editor. Memo 62, Univ. Edinburgh, Dept. of Computational Logic.

[Boyer & Moore, 1971] Robert S. Boyer and J Strother Moore, 1971. The sharing of structure in resolution programs. Memo 47, Univ. Edinburgh, Dept. of Computational Logic. II + 24 pp.. Revised version is [Boyer & Moore, 1972].

[Boyer & Moore, 1972] Robert S. Boyer and J Strother Moore, 1972. The sharing of structure in theorem-proving programs. In [Meltzer & Michie, 1972, pp. 101–116].

[Boyer & Moore, 1973] Robert S. Boyer and J Strother Moore, 1973. Proving theorems about LISP functions. In [Nilsson, 1973, pp. 486–493]. http://ijcai.org/Past%20Proceedings/IJCAI-73/PDF/053.pdf. Rev. version, extd. with a section "Failures", is [Boyer & Moore, 1975].

[Boyer & Moore, 1975] Robert S. Boyer and J Strother Moore, 1975. Proving theorems about LISP functions. *J. of the ACM*, 22:129–144. Rev. extd. edn. of [Boyer & Moore, 1973]. Received Sept. 1973, Rev. April 1974.

[Boyer & Moore, 1977] Robert S. Boyer and J Strother Moore, 1977. A fast string searching algorithm. *Comm. ACM*, 20:762–772. http://doi.acm.org/10.1145/359842.359859.

[Boyer & Moore, 1977] Robert S. Boyer and J Strother Moore, 1977. A lemma driven automatic theorem prover for recursive function theory. In [Reddy, 1977, Vol. I, pp. 511–519]. http://ijcai.org/Past%20Proceedings/IJCAI-77-VOL1/PDF/089.pdf.

[Boyer & Moore, 1979] Robert S. Boyer and J Strother Moore, 1979. *A Computational Logic*. Academic Press (Elsevier). http://www.cs.utexas.edu/users/boyer/acl.text.

[Boyer & Moore, 1981a] Robert S. Boyer and J Strother Moore, editors, 1981a. *The Correctness Problem in Computer Science*. Academic Press (Elsevier).

[Boyer & Moore, 1981b] Robert S. Boyer and J Strother Moore, 1981b. Metafunctions: Proving them correct and using them efficiently as new proof procedures. In [Boyer & Moore, 1981a, pp. 103–184].

[Boyer & Moore, 1984a] Robert S. Boyer and J Strother Moore, 1984a. A mechanical proof of the Turing completeness of pure LISP. In [Bledsoe & Loveland, 1984, pp. 133–167].

[Boyer & Moore, 1984b] Robert S. Boyer and J Strother Moore, 1984b. A mechanical proof of the unsolvability of the halting problem. *J. of the ACM*, 31:441–458.

[Boyer & Moore, 1984c] Robert S. Boyer and J Strother Moore, 1984c. Proof checking the RSA public key encryption algorithm. *American Mathematical Monthly*, 91:181–189.

[Boyer & Moore, 1985] Robert S. Boyer and J Strother Moore, 1985. Program verification. *J. Automated Reasoning*, 1:17–23.

[Boyer & Moore, 1987] Robert S. Boyer and J Strother Moore, 1987. The addition of bounded quantification and partial functions to a computational logic and its theorem prover. Technical Report ICSCA-CMP-52, Inst. for Computing Science and Computing Applications, The University of Texas at Austin. Printed Jan. 1987. Also published as [Boyer & Moore, 1988a; 1989].

[Boyer & Moore, 1988a] Robert S. Boyer and J Strother Moore, 1988a. The addition of bounded quantification and partial functions to a computational logic and its theorem prover. *J. Automated Reasoning*, 4:117–172. Received Feb. 11, 1987. Also pubished as [Boyer & Moore, 1987; 1989].

[Boyer & Moore, 1988b] Robert S. Boyer and J Strother Moore, 1988b. *A Computational Logic Handbook*. Number 23 in Perspectives in Computing. Academic Press (Elsevier). 2nd rev. extd. edn. is [Boyer & Moore, 1998].

[Boyer & Moore, 1988c] Robert S. Boyer and J Strother Moore, 1988c. Integrating decision procedures into heuristic theorem provers: A case study of linear arithmetic. In [Hayes &al., 1988, pp. 83–124].

[Boyer & Moore, 1989] Robert S. Boyer and J Strother Moore, 1989. The addition of bounded quantification and partial functions to a computational logic and its theorem prover. In [Broy, 1989, pp. 95–145] (received Jan. 1988). Also published as [Boyer & Moore, 1987; 1988a].

[Boyer & Moore, 1990] Robert S. Boyer and J Strother Moore, 1990. A theorem prover for a computational logic. In [Stickel, 1990, pp. 1–15].

[Boyer & Moore, 1998] Robert S. Boyer and J Strother Moore, 1998. *A Computational Logic Handbook*. International Series in Formal Methods. Academic Press (Elsevier). 2nd rev. extd. edn. of [Boyer & Moore, 1988b], rev. to work with NQTHM–1992, a new version of NQTHM.

[Boyer &al., 1976] Robert S. Boyer, J Strother Moore, and Robert E. Shostak, 1976. Primitive recursive program transformations. In [Graham &al., 1976, pp. 171–174]. http://doi.acm.org/10.1145/800168.811550.

[Boyer & Yu, 1992] Robert S. Boyer and Yuan Yu, 1992. Automated correctness proofs of machine code programs for a commercial microprocessor. In [Kapur, 1992, 416–430].

[Boyer & Yu, 1996] Robert S. Boyer and Yuan Yu, 1996. Automated proofs of object code for a widely used microprocessor. J. of the ACM, 43:166–192.

[Brock & Hunt, 1999] Bishop Brock and Warren A. Hunt, 1999. Formal analysis of the Motorola CAP DSP. In [Hinchey & Bowen, 1999, pp. 81–116].

[Brotherston, 2005] James Brotherston, 2005. Cyclic proofs for first-order logic with inductive definitions. In [Beckert, 2005, pp. 78–92]. Thoroughly rev. version in [Brotherston & Simpson, 2011].

[Brotherston & Simpson, 2007] James Brotherston and Alex Simpson, 2007. Complete sequent calculi for induction and infinite descent. In [LICS, 2007, pp. 51–62?]. Thoroughly rev. version in [Brotherston & Simpson, 2011].

[Brotherston & Simpson, 2011] James Brotherston and Alex Simpson, 2011. Sequent calculi for induction and infinite descent. *J. Logic and Computation*, 21:1177–1216. Thoroughly rev. version of [Brotherston, 2005] and [Brotherston & Simpson, 2007]. Received April 3, 2009. Published online Sept. 30, 2010, http://dx.doi.org/10.1093/logcom/exq052.

[Brown, 2012] Chad E. Brown, 2012. SATALLAX: An automatic higher-order prover. In [Gramlich &al., 2012, pp. 111–117].

[Broy, 1989] Manfred Broy, editor, 1989. *Constructive Methods in Computing Science*, number F 55 in NATO ASI Series. Springer.

[Buch & Hillenbrand, 1996] Armin Buch and Thomas Hillenbrand, 1996. WALDMEISTER: *Development of a High Performance Completion-Based Theorem Prover*. SEKI-Report SR–96–01 (ISSN 1860–5931). SEKI Publications, FB Informatik, Univ. Kaiserslautern. agent.informatik.uni-kl.de/seki/1996/Buch.SR-96-01.ps.gz.

[Bundy, 1988] Alan Bundy, 1988. *The use of Explicit Plans to Guide Inductive Proofs*. DAI Research Paper No. 349, Dept. Artificial Intelligence, Univ. Edinburgh. Short version in [Lusk & Overbeek, 1988, pp. 111–120].

[Bundy, 1989] Alan Bundy, 1989. *A Science of Reasoning*. DAI Research Paper No. 445, Dept. Artificial Intelligence, Univ. Edinburgh. Also in [Lassez & Plotkin, 1991, pp. 178–198].

[Bundy, 1994] Alan Bundy, editor, 1994. *12th Int. Conf. on Automated Deduction (CADE), Nancy, 1994*, number 814 in Lecture Notes in Artificial Intelligence. Springer.

[Bundy, 1999] Alan Bundy, 1999. *The Automation of Proof by Mathematical Induction*. Informatics Research Report No. 2, Division of Informatics, Univ. Edinburgh. Also in [Robinson & Voronkow, 2001, Vol. 1, pp. 845–911].

[Bundy &al., 1989] Alan Bundy, Frank van Harmelen, Jane Hesketh, Alan Smaill, and Andrew Stevens, 1989. A rational reconstruction and extension of recursion analysis. In [Sridharan, 1989, pp. 359–365].

[Bundy &al., 1990] Alan Bundy, Frank van Harmelen, Christian Horn, and Alan Smaill, 1990. The OYSTER/CLAM system. In [Stickel, 1990, pp. 647–648].

[Bundy &al., 2005] Alan Bundy, Dieter Hutter, David Basin, and Andrew Ireland, 2005. *Rippling: Meta-Level Guidance for Mathematical Reasoning*. Cambridge Univ. Press.

[Bundy &al., 1991] Alan Bundy, Andrew Stevens, Frank van Harmelen, Andrew Ireland, and Alan Smaill, 1991. *Rippling: A Heuristic for Guiding Inductive Proofs*. DAI Research Paper No. 567, Dept. Artificial Intelligence, Univ. Edinburgh. Also in Artificial Intelligence 62:185–253, 1993.

[Burstall, 1969] Rod M. Burstall, 1969. Proving properties of programs by structural induction. *The Computer Journal*, 12:48–51. Received April 1968, rev. Aug. 1968.

[Burstall &al., 1971] Rod M. Burstall, John S. Collins, and Robin J. Popplestone, 1971. *Programming in* POP-2. Univ. Edinburgh Press.

[Bussey, 1917] W. H. Bussey, 1917. The origin of mathematical induction. *American Mathematical Monthly*, XXIV:199–207.

[Bussotti, 2006] Paolo Bussotti, 2006. *From Fermat to Gauß: indefinite descent and methods of reduction in number theory*. Number 55 in Algorismus. Dr. Erwin Rauner Verlag, Augsburg.

[Cajori, 1918] Florian Cajori, 1918. Origin of the name "mathematical induction". *American Mathematical Monthly*, 25:197–201.

[Church, 1946] Alonzo Church, 1946. Review of [Bourbaki, 1939]. *J. Symbolic Logic*, 11:91.

[Clocksin & Mellish, 2003] William F. Clocksin and Christopher S. Mellish, 2003. *Programming in* PROLOG. Springer. 5th edn. (1st edn. 1981).

[Cohn, 1965] Paul Moritz Cohn, 1965. *Universal Algebra*. Harper & Row, New York. 1st edn.. 2nd rev. edn. is [Cohn, 1981].

[Cohn, 1981] Paul Moritz Cohn, 1981. *Universal Algebra*. Number 6 in Mathematics and Its Applications. D. Reidel Publ. (Springer Science+Business Media), Dordrecht (The Netherlands). 2nd rev. edn. (1st edn. is [Cohn, 1965]).

[Comon, 1997] Hubert Comon, editor, 1997. *8th Int. Conf. on Rewriting Techniques and Applications (RTA), Sitges (Spain), 1997*, number 1232 in Lecture Notes in Computer Science. Springer.

[Comon, 2001] Hubert Comon, 2001. Inductionless induction. In [Robinson & Voronkow, 2001, Vol. I, pp. 913–970].

[Constable &al., 1985] Robert L. Constable, Stuart F. Allen, H. M. Bromly, W. R. Cleaveland, J. F. Cremer, R. W. Harper, Douglas J. Howe, T. B. Knoblock, N. P. Mendler, P. Panangaden, James T. Sasaki, and Scott F. Smith, 1985. *Implementing Mathematics with the* NUPRL *Proof Development System*. Prentice–Hall, Inc.. http://www.nuprl.org/book.

[Cooper, 1971] D. C. Cooper, editor, 1971. *Proc. 2nd Int. Joint Conf. on Artificial Intelligence (IJCAI), Sept. 1971, Imperial College, London*. Morgan Kaufmann, Los Altos (CA), Los Altos (CA). http://ijcai.org/Past%20Proceedings/IJCAI-1971/

CONTENT/content.htm.

[DAC, 2001] DAC, 2001. *Proc. 38th Design Automation Conference (DAC), Las Vegas (NV), 2001*. ACM Press.

[Darlington, 1968] Jared L. Darlington, 1968. Automated theorem proving with equality substitutions and mathematical induction. In [Michie, 1968, pp. 113–127].

[Davis, 2009] Jared Davis, 2009. *A Self-Verifying Theorem Prover*. PhD thesis, The University of Texas at Austin.

[Dedekind, 1888] Richard Dedekind, 1888. *Was sind und was sollen die Zahlen?*. Verlag von Friedrich Vieweg und Sohn, Braunschweig. Also in [Dedekind, 1930–32, Vol. 3, pp. 335–391]. Also in [Dedekind, 1969].

[Dedekind, 1930–32] Richard Dedekind, 1930–32. *Gesammelte mathematische Werke*. Verlag von Friedrich Vieweg und Sohn, Braunschweig. Ed. by Robert Fricke, Emmy Noether, and Öystein Ore.

[Dedekind, 1969] Richard Dedekind, 1969. *Was sind und was sollen die Zahlen? Stetigkeit und irrationale Zahlen*. Verlag von Friedrich Vieweg und Sohn, Braunschweig.

[Dennis &al., 2005] Louise A. Dennis, Mateja Jamnik, and Martin Pollet, 2005. On the comparison of proof planning systems λCLAM, ΩMEGA and IsaPlanner. *Electronic Notes in Theoretical Computer Sci.*, 151:93–110.

[Dershowitz, 1989] Nachum Dershowitz, editor, 1989. *3rd Int. Conf. on Rewriting Techniques and Applications (RTA), Chapel Hill (NC), 1989*, number 355 in Lecture Notes in Computer Science. Springer.

[Dershowitz & Jouannaud, 1990] Nachum Dershowitz and Jean-Pierre Jouannaud, 1990. Rewrite systems. In [Leeuwen, 1990, Vol. B, pp. 243–320].

[Dershowitz & Lindenstrauss, 1995] Nachum Dershowitz and Naomi Lindenstrauss, editors, 1995. *4th Int. Workshop on Conditional Term Rewriting Systems (CTRS), Jerusalem, 1994*, number 968 in Lecture Notes in Computer Science.

[Dietrich, 2011] Dominik Dietrich, 2011. *Assertion Level Proof Planning with Compiled Strategies*. Optimus Verlag, Alexander Mostafa, Göttingen. PhD thesis, Dept. Informatics, FR Informatik, Saarland Univ..

[Dixon & Fleuriot, 2003] Lucas Dixon and Jacques Fleuriot, 2003. IsaPlanner: A prototype proof planner in Isabelle. In [Baader, 2003, pp. 279–283].

[Eisenreich & Sube, 1982] Günther Eisenreich and Ralf Sube, 1982. *Wörterbuch der Mathematik: Englisch–Deutsch–Französisch–Russisch*. VEB Verlag Technik, Berlin. Two volumes, also under license to Verlag Harri Deutsch, Thun (Switzerland), 1982. Reprinted in one volume as *Langenscheidts Fachwörterbuch Mathematik* by Langenscheidt, Berlin, 1996.

[Euclid, ca. 300 B.C.] Euclid of Alexandria, ca. 300 B.C.. *Elements*. Web version without the figures: http://www.perseus.tufts.edu/hopper/text?doc=Perseus:text:1999.01.0085. English translation: Thomas L. Heath (ed.). *The Thirteen Books of Euclid's Elements*. Cambridge Univ. Press, 1908; web version without the figures: http://www.perseus.tufts.edu/hopper/text?doc=Perseus:text:1999.01.0086. English web version (incl. figures): D. E. Joyce (ed.). *Euclid's Elements*. http://aleph0.clarku.edu/~djoyce/java/elements/elements.html, Dept. Math. & Comp. Sci., Clark Univ., Worcester (MA).

[Fermat, 1891ff.] Pierre Fermat, 1891ff.. *Œuvres de Fermat*. Gauthier-Villars, Paris. Ed. by Paul Tannery, Charles Henry.

[Fitting, 1990] Melvin Fitting, 1990. *First-order logic and automated theorem proving*.

Springer. 1st edn. (2nd rev. edn. is [Fitting, 1996]).

[Fitting, 1996] Melvin Fitting, 1996. *First-order logic and automated theorem proving*. Springer. 2nd rev. edn. (1st edn. is [Fitting, 1990]).

[FOCS, 1980] FOCS, 1980. *Proc. 21st Annual Symposium on Foundations of Computer Sci., Syracuse, 1980*. IEEE Press. http://ieee-focs.org/.

[Fowler, 1994] David Fowler, 1994. Could the Greeks have used mathematical induction? Did they use it? *Physis*, XXXI(1):253–265.

[Freudenthal, 1953] Hans Freudenthal, 1953. Zur Geschichte der vollständigen Induktion. *Archives Internationales d'Histoire des Sciences*, 6:17–37.

[Fribourg, 1986] Laurent Fribourg, 1986. A strong restriction of the inductive completion procedure. In [Kott, 1986, pp. 105–116]. Also in J. Symbolic Computation 8:253–276, 1989, Academic Press (Elsevier).

[Fries, 1822] Jakob Friedrich Fries, 1822. *Die mathematische Naturphilosophie nach philosophischer Methode bearbeitet – Ein Versuch*. Christian Friedrich Winter, Heidelberg. Facsimlie in [Fries, 1967ff., Vol. 13 (1979)].

[Fries, 1967ff.] Jakob Friedrich Fries, 1967ff.. *Sämtliche Schriften*. Scientia Verlag (Kloof Booksellers & Scientia Verlag), Aalen (Germany). 33 volumes, ed. by GERT KÖNIG and LUTZ GELDSETZER.

[Fritz, 1945] Kurt von Fritz, 1945. The discovery of incommensurability by Hippasus of Metapontum. *Annals of Mathematics*, 46:242–264. German translation: *Die Entdeckung der Inkommensurabilität durch Hippasos von Metapont* in [Becker, 1965, pp. 271–308].

[Fuchi & Kott, 1988] Kazuhiro Fuchi and Laurent Kott, editors, 1988. *Programming of Future Generation Computers II: Proc. of the 2nd Franco-Japanese Symposium*. North-Holland (Elsevier).

[Gabbay &al., 1994] Dov Gabbay, Christopher John Hogger, and J. Alan Robinson, editors, 1994. *Handbook of Logic in Artificial Intelligence and Logic Programming. Vol. 2: Deduction Methodologies*. Oxford Univ. Press.

[Gabbay & Woods, 2004ff.] Dov Gabbay and John Woods, editors, 2004ff.. *Handbook of the History of Logic*. North-Holland (Elsevier).

[Ganzinger, 1996] Harald Ganzinger, editor, 1996. *7th Int. Conf. on Rewriting Techniques and Applications (RTA), New Brunswick (NJ), 1996*, number 1103 in Lecture Notes in Computer Science. Springer.

[Ganzinger, 1999] Harald Ganzinger, editor, 1999. *16th Int. Conf. on Automated Deduction (CADE), Trento (Italy), 1999*, number 1632 in Lecture Notes in Artificial Intelligence. Springer.

[Ganzinger & Stuber, 1992] Harald Ganzinger and Jürgen Stuber, 1992. Inductive Theorem Proving by Consistency for First-Order Clauses. In [Bachmair &al., 1992, pp. 441–462]. Also in [Rusinowitch & Remy, 1993, pp. 226–241].

[Gentzen, 1935] Gerhard Gentzen, 1935. Untersuchungen über das logische Schließen. *Mathematische Zeitschrift*, 39:176–210,405–431. Also in [Berka & Kreiser, 1973, pp. 192–253]. English translation in [Gentzen, 1969].

[Gentzen, 1969] Gerhard Gentzen, 1969. *The Collected Papers of Gerhard Gentzen*. North-Holland (Elsevier). Ed. by Manfred E. Szabo.

[Geser, 1995] Alfons Geser, 1995. A principle of non-wellfounded induction. In [Margaria,

1995, pp. 117–124].

[Geser, 1996] Alfons Geser, 1996. An improved general path order. *J. Applicable Algebra in Engineering, Communication and Computing (AAECC)*, 7:469–511.

[Gillman, 1987] Leonard Gillman, 1987. *Writing Mathematics Well*. The Mathematical Association of America.

[Göbel, 1985] Richard Göbel, 1985. Completion of globally finite term rewriting systems for inductive proofs. In [Stoyan, 1985, pp. 101–110].

[Gödel, 1931] Kurt Gödel, 1931. Über formal unentscheidbare Sätze der Principia Mathematica und verwandter Systeme I. *Monatshefte für Mathematik und Physik*, 38:173–198. With English translation also in [Gödel, 1986ff., Vol. I, pp. 145–195]. English translation also in [Heijenoort, 1971, pp. 596–616] and in [Gödel, 1962].

[Gödel, 1962] Kurt Gödel, 1962. *On formally undecidable propositions of Principia Mathematica and related systems*. Basic Books, New York. English translation of [Gödel, 1931] by Bernard Meltzer. With an introduction by R. B. Braithwaite. 2nd edn. by Dover Publications, 1992.

[Gödel, 1986ff.] Kurt Gödel, 1986ff. *Collected Works*. Oxford Univ. Press. Ed. by Sol Feferman, John W. Dawson Jr., Warren Goldfarb, Jean van Heijenoort, Stephen C. Kleene, Charles Parsons, Wilfried Sieg, *et al.*.

[Goguen, 1980] Joseph Goguen, 1980. How to prove algebraic inductive hypotheses without induction. In [Bibel & Kowalski, 1980, pp. 356–373].

[Goldstein, 2008] Catherine Goldstein, 2008. Pierre Fermat. In [Gowers &al., 2008, § VI.12, pp. 740–741].

[Goodstein, 1945] R. L. Goodstein, 1945. Function theory in an axiom-free equation calculus. *Proceedings of the London Mathematical Society, Ser. 2*, 48:401–434.

[Goodstein, 1957] R. L. Goodstein, 1957. *Recursive number theory — A development of recursive arithmetic in a logic-free equation calculus*. Studies in logic and the foundations of mathematics. North-Holland (Elsevier), Amsterdam. 2nd edn. 1965.

[Gordon, 2000] Mike J. C. Gordon, 2000. From LCF to HOL: a short history. In [Plotkin &al., 2000, pp. 169–186]. http://www.cl.cam.ac.uk/~mjcg/papers/HolHistory.pdf.

[Gore &al., 2001] Rajeev Gore, Alexander Leitsch, and Tobias Nipkow, editors, 2001. *1st Int. Joint Conf. on Automated Reasoning (IJCAR), Siena (Italy), 2001*, number 2083 in Lecture Notes in Artificial Intelligence. Springer.

[Gowers &al., 2008] Timothy Gowers, June Barrow-Green, and Imre Leader, editors, 2008. *The Princeton Companion to Mathematics*. Princeton Univ. Press.

[Graham &al., 1976] Susan L. Graham, Robert M. Graham, Michael A. Harrison, William I. Grosky, and Jeffrey D. Ullman, editors, 1976. *Conference Record of the 3rd Annual ACM SIGPLAN-SIGACT Symposium on Principles of Programming Languages (POPL), Atlanta (GA), Jan. 1976*. ACM Press. http://dl.acm.org/citation.cfm?id=800168.

[Gramlich, 1996] Bernhard Gramlich, 1996. *Termination and Confluence Properties of Structured Rewrite Systems*. PhD thesis, FB Informatik, Univ. Kaiserslautern. www.logic.at/staff/gramlich/papers/thesis96.pdf. x+217 pp..

[Gramlich & Lindner, 1991] Bernhard Gramlich and Wolfgang Lindner, 1991. *A Guide to UNICOM, an Inductive Theorem Prover Based on Rewriting and Completion Techniques*. SEKI-Report SR–91–17 (ISSN 1860–5931). SEKI Publications, FB Informatik,

Univ. Kaiserslautern. http://agent.informatik.uni-kl.de/seki/1991/Lindner.
SR-91-17.ps.gz.

[Gramlich &al., 2012] Bernhard Gramlich, Dale A. Miller, and Uli Sattler, editors, 2012. *6th Int. Joint Conf. on Automated Reasoning (IJCAR), Manchester, 2012*, number 7364 in Lecture Notes in Artificial Intelligence. Springer.

[Gramlich & Wirth, 1996] Bernhard Gramlich and Claus-Peter Wirth, 1996. Confluence of terminating conditional term rewriting systems revisited. In [Ganzinger, 1996, pp. 245–259].

[Hayes &al., 1988] Jean E. Hayes, Donald Michie, and Judith Richards, editors, 1988. *Proceedings of the 11th Annual Machine Intelligence Workshop (Machine Intelligence 11), Univ. Strathclyde, Glasgow, 1985*. Clarendon Press (Oxford Univ. Press), Oxford. aitopics.org/sites/default/files/classic/Machine_Intelligence_11/Machine_Intelligence_v.11.pdf.

[Heijenoort, 1971] Jean van Heijenoort, 1971. *From Frege to Gödel: A Source Book in Mathematical Logic, 1879–1931*. Harvard Univ. Press. 2nd rev. edn. (1st edn. 1967).

[Herbelin, 2009] Hugo Herbelin, editor, 2009. *The 1st Coq Workshop*. Inst. für Informatik, Tech. Univ. München. TUM-I0919, http://www.lix.polytechnique.fr/coq/files/coq-workshop-TUM-I0919.pdf.

[Hilbert, 1899] David Hilbert, 1899. Grundlagen der Geometrie. In [Anon, 1899, pp. 1–92]. 1st edn. without appendixes. Reprinted in [Hilbert, 2004, pp. 436–525]. *(Last edition of "Grundlagen der Geometrie" by Hilbert is [Hilbert, 1930b], which is also most complete regarding the appendixes. Last three editions by Paul Bernays are [Hilbert, 1962; 1968; 1972], which are also most complete regarding supplements and figures. Its first appearance as a separate book was the French translation [Hilbert, 1900b]. Two substantially different English translations are [Hilbert, 1902] and [Hilbert, 1971]).*

[Hilbert, 1900a] David Hilbert, 1900a. Über den Zahlbegriff. *Jahresbericht der Deutschen Mathematiker-Vereinigung*, 8:180–184. Received Dec. 1899. Reprinted as Appendix VI of [Hilbert, 1909; 1913; 1922; 1923; 1930b].

[Hilbert, 1900b] David Hilbert, 1900b. Les principes fondamentaux de la géométrie. *Annales Scientifiques de l'École Normale Supérieure*, Série 3, 17:103–209. French translation by Léonce Laugel of special version of [Hilbert, 1899], revised and authorized by Hilbert. Also in published as a separate book by the same publisher (Gauthier-Villars, Paris).

[Hilbert, 1902] David Hilbert, 1902. *The Foundations of Geometry*. Open Court, Chicago. English translation by E. J. Townsend of special version of [Hilbert, 1899], revised and authorized by Hilbert, http://www.gutenberg.org/etext/17384.

[Hilbert, 1903] David Hilbert, 1903. *Grundlagen der Geometrie. — Zweite, durch Zusätze vermehrte und mit fünf Anhängen versehene Auflage. Mit zahlreichen in den Text gedruckten Figuren*. Druck und Verlag von B. G. Teubner, Leipzig. 2nd rev. extd. edn. of [Hilbert, 1899], rev. and extd. with five appendixes, newly added figures, and an index of notion names.

[Hilbert, 1905] David Hilbert, 1905. Über die Grundlagen der Logik und der Arithmetik. In [Krazer, 1905, pp. 174–185]. Reprinted as Appendix VII of [Hilbert, 1909; 1913; 1922; 1923; 1930b]. English translation *On the foundations of logic and arithmetic* by Beverly Woodward with an introduction by Jean van Heijenoort in [Heijenoort, 1971, pp. 129–138].

[Hilbert, 1909] David Hilbert, 1909. *Grundlagen der Geometrie. — Dritte, durch Zusätze* und Literaturhinweise von neuem vermehrte und mit sieben Anhängen versehene Auflage. Mit zahlreichen in den Text gedruckten Figuren. Number VII in Wissenschaft und Hypothese. Druck und Verlag von B. G. Teubner, Leipzig, Berlin. 3rd rev. extd. edn. of [Hilbert, 1899], rev. edn. of [Hilbert, 1903], extd. with a bibliography and two additional appendixes (now seven in total) (Appendix VI: [Hilbert, 1900a]) (Appendix VII: [Hilbert, 1905]).

[Hilbert, 1913] David Hilbert, 1913. *Grundlagen der Geometrie. — Vierte, durch Zusätze und Literaturhinweise von neuem vermehrte und mit sieben Anhängen versehene Auflage. Mit zahlreichen in den Text gedruckten Figuren*. Druck und Verlag von B. G. Teubner, Leipzig, Berlin. 4th rev. extd. edn. of [Hilbert, 1899], rev. edn. of [Hilbert, 1909].

[Hilbert, 1922] David Hilbert, 1922. *Grundlagen der Geometrie. — Fünfte, durch Zusätze und Literaturhinweise von neuem vermehrte und mit sieben Anhängen versehene Auflage. Mit zahlreichen in den Text gedruckten Figuren*. Verlag und Druck von B. G. Teubner, Leipzig, Berlin. 5th extd. edn. of [Hilbert, 1899]. Contrary to what the subtitle may suggest, this is an anastatic reprint of [Hilbert, 1913], extended by a very short preface on the changes w.r.t. [Hilbert, 1913], and with augmentations to Appendix II, Appendix III, and Chapter IV, § 21.

[Hilbert, 1923] David Hilbert, 1923. *Grundlagen der Geometrie. — Sechste unveränderte Auflage. Anastatischer Nachdruck. Mit zahlreichen in den Text gedruckten Figuren*. Verlag und Druck von B. G. Teubner, Leipzig, Berlin. 6th rev. extd. edn. of [Hilbert, 1899], anastatic reprint of [Hilbert, 1922].

[Hilbert, 1926] David Hilbert, 1926. Über das Unendliche — Vortrag, gehalten am 4. Juni 1925 gelegentlich einer zur Ehrung des Andenkens an Weierstraß von der Westfälischen Math. Ges. veranstalteten Mathematiker-Zusammenkunft in Münster i. W. *Mathematische Annalen*, 95:161–190. Received June 24, 1925. Reprinted as Appendix VIII of [Hilbert, 1930b]. English translation *On the infinite* by Stefan Bauer-Mengelberg with an introduction by Jean van Heijenoort in [Heijenoort, 1971, pp. 367–392].

[Hilbert, 1928] David Hilbert, 1928. Die Grundlagen der Mathematik — Vortrag, gehalten auf Einladung des Mathematischen Seminars im Juli 1927 in Hamburg. *Abhandlungen aus dem mathematischen Seminar der Univ. Hamburg*, 6:65–85. Reprinted as Appendix IX of [Hilbert, 1930b]. English translation *The foundations of mathematics* by Stefan Bauer-Mengelberg and Dagfinn Føllesdal with a short introduction by Jean van Heijenoort in [Heijenoort, 1971, pp. 464–479].

[Hilbert, 1930a] David Hilbert, 1930a. Probleme der Grundlegung der Mathematik. *Mathematische Annalen*, 102:1–9. Vortrag gehalten auf dem Internationalen Mathematiker-Kongreß in Bologna, Sept. 3, 1928. Received March 25, 1929. Reprinted as Appendix X of [Hilbert, 1930b]. Short version in *Atti del congresso internazionale dei matematici, Bologna, 3–10 settembre 1928*, Vol. 1, pp. 135–141, Bologna, 1929.

[Hilbert, 1930b] David Hilbert, 1930b. *Grundlagen der Geometrie. — Siebente umgearbeitete und vermehrte Auflage. Mit 100 in den Text gedruckten Figuren*. Verlag und Druck von B. G. Teubner, Leipzig, Berlin. 7th rev. extd. edn. of [Hilbert, 1899], thoroughly revised edition of [Hilbert, 1923], extd. with three new appendixes (now ten in total) (Appendix VIII: [Hilbert, 1926]) (Appendix IX: [Hilbert, 1928]) (Appendix X: [Hilbert, 1930a]).

[Hilbert, 1956] David Hilbert, 1956. *Grundlagen der Geometrie. — Achte Auflage, mit*

[Hilbert, 1962] *Revisionen und Ergänzungen von Dr. Paul Bernays. Mit 124 Abbildungen.* B. G. Teubner Verlagsgesellschaft, Stuttgart. 8th rev. extd. edn. of [Hilbert, 1899], rev. edn. of [Hilbert, 1930b], omitting appendixes VI–X, extd. by Paul Bernays, now with 24 additional figures and 3 additional supplements.

[Hilbert, 1962] David Hilbert, 1962. *Grundlagen der Geometrie. — Neunte Auflage, revidiert und ergänzt von Dr. Paul Bernays. Mit 129 Abbildungen.* B. G. Teubner Verlagsgesellschaft, Stuttgart. 9th rev. extd. edn. of [Hilbert, 1899], rev. edn. of [Hilbert, 1956], extd. by Paul Bernays, now with 129 figures, 5 appendixes, and 8 supplements (I 1, I 2, II, III, IV 1, IV 2, V 1, V 2).

[Hilbert, 1968] David Hilbert, 1968. *Grundlagen der Geometrie. — Zehnte Auflage, revidiert und ergänzt von Dr. Paul Bernays. Mit 124 Abbildungen.* B. G. Teubner Verlagsgesellschaft, Stuttgart. 10th rev. extd. edn. of [Hilbert, 1899], rev. edn. of [Hilbert, 1962] by Paul Bernays.

[Hilbert, 1971] David Hilbert, 1971. *The Foundations of Geometry.* Open Court, Chicago and La Salle (IL). Newly translated and fundamentally different 2nd edn. of [Hilbert, 1902], actually an English translation of [Hilbert, 1968] by Leo Unger.

[Hilbert, 1972] David Hilbert, 1972. *Grundlagen der Geometrie. — 11. Auflage. Mit Supplementen von Dr. Paul Bernays.* B. G. Teubner Verlagsgesellschaft, Stuttgart. 11th rev. extd. edn. of [Hilbert, 1899], rev. edn. of [Hilbert, 1968] by Paul Bernays.

[Hilbert, 2004] David Hilbert, 2004. *David Hilbert's Lectures on the Foundations of Geometry, 1891–1902.* Springer. Ed. by Michael Hallett and Ulrich Majer.

[Hilbert & Bernays, 1934] David Hilbert and Paul Bernays, 1934. *Grundlagen der Mathematik — Erster Band.* Number XL in Grundlehren der mathematischen Wissenschaften. Springer. 1st edn. (2nd edn. is [Hilbert & Bernays, 1968]). Reprinted by J. W. Edwards Publ., Ann Arbor (MI), 1944. English translation is [Hilbert & Bernays, 2017a; 2017b].

[Hilbert & Bernays, 1939] David Hilbert and Paul Bernays, 1939. *Grundlagen der Mathematik — Zweiter Band.* Number L in Grundlehren der mathematischen Wissenschaften. Springer. 1st edn. (2nd edn. is [Hilbert & Bernays, 1970]). Reprinted by J. W. Edwards Publ., Ann Arbor (MI), 1944.

[Hilbert & Bernays, 1968] David Hilbert and Paul Bernays, 1968. *Grundlagen der Mathematik I.* Number 40 in Grundlehren der mathematischen Wissenschaften. Springer. 2nd rev. edn. of [Hilbert & Bernays, 1934]. English translation is [Hilbert & Bernays, 2017a; 2017b].

[Hilbert & Bernays, 1970] David Hilbert and Paul Bernays, 1970. *Grundlagen der Mathematik II.* Number 50 in Grundlehren der mathematischen Wissenschaften. Springer. 2nd rev. extd. edn. of [Hilbert & Bernays, 1939].

[Hilbert & Bernays, 2017a] David Hilbert and Paul Bernays, 2017a. *Grundlagen der Mathematik I — Foundations of Mathematics I, Part A: Title Pages, Prefaces, and §§ 1–2.* Springer. First English translation and bilingual facsimile edn. of the 2nd German edn. [Hilbert & Bernays, 1968], incl. the annotation and translation of all differences of the 1st German edn. [Hilbert & Bernays, 1934]. Ed. by Claus-Peter Wirth, Jörg Siekmann, Volker Peckhaus, Michael Gabbay, Dov Gabbay. Translated and commented by Claus-Peter Wirth *et al.* Thoroughly rev. 3rd edn. (1st edn. College Publications, London, 2011; 2nd edn. http://wirth.bplaced.net/p/hilbertbernays, 2013).

[Hilbert & Bernays, 2017b] David Hilbert and Paul Bernays, 2017b. *Grundlagen der*

Mathematik I — Foundations of Mathematics I, Part B: §§ 3–5 and Deleted Part 1 of the 1ˢᵗ Edn.. Springer. First English translation and bilingual facsimile edn. of the 2ⁿᵈ German edn. [Hilbert & Bernays, 1968], incl. the annotation and translation of all deleted texts of the 1ˢᵗ German edn. [Hilbert & Bernays, 1934]. Ed. by Claus-Peter Wirth, Jörg Siekmann, Volker Peckhaus, Michael Gabbay, Dov Gabbay. Translated and commented by Claus-Peter Wirth *et al.* Thoroughly rev. 3ʳᵈ edn. (1ˢᵗ edn. College Publications, London, 2012; 2ⁿᵈ edn. http://wirth.bplaced.net/p/hilbertbernays, 2013).

[Hillenbrand & Löchner, 2002] Thomas Hillenbrand and Bernd Löchner, 2002. The next WALDMEISTER loop. In [Voronkov, 2002, pp. 486–500]. http://www.waldmeister.org.

[Hinchey & Bowen, 1999] Michael G. Hinchey and Jonathan P. Bowen, editors, 1999. *Industrial-Strength Formal Methods in Practice*. Formal Approaches to Computing and Information Technology (FACIT). Springer.

[Hobson & Love, 1913] E. W. Hobson and A. E. H. Love, editors, 1913. *Proc. 5ᵗʰ Int. Congress of Mathematicians, Cambridge, Aug 22–28, 1912*. Cambridge Univ. Press. http://gallica.bnf.fr/ark:/12148/bpt6k99444q.

[Howard & Rubin, 1998] Paul Howard and Jean E. Rubin, 1998. *Consequences of the Axiom of Choice*. American Math. Soc.. http://consequences.emich.edu/conseq.htm.

[Hudlak &al., 1999] Paul Hudlak, John Peterson, and Joseph H. Fasel, 1999. A gentle introduction to HASKELL. Web only: http://www.haskell.org/tutorial.

[Huet, 1980] Gérard Huet, 1980. Confluent reductions: Abstract properties and applications to term rewriting systems. *J. of the ACM*, 27:797–821.

[Huet & Hullot, 1980] Gérard Huet and Jean-Marie Hullot, 1980. Proofs by induction in equational theories with constructors. In [FOCS, 1980, pp. 96–107]. Also in J. Computer and System Sci. 25:239–266, 1982, Academic Press (Elsevier).

[Hunt, 1985] Warren A. Hunt, 1985. *FM8501: A Verified Microprocessor*. PhD thesis, The University of Texas at Austin. Also published as [Hunt, 1994].

[Hunt, 1989] Warren A. Hunt, 1989. Microprocessor design verification. *J. Automated Reasoning*, 5:429–460.

[Hunt, 1994] Warren A. Hunt, 1994. *FM8501: A Verified Microprocessor*. Number 795 in Lecture Notes in Artificial Intelligence. Springer. Originally published as [Hunt, 1985].

[Hunt & Swords, 2009] Warren A. Hunt and Sol Swords, 2009. Centaur technology media unit verification. In [Bouajjani & Maler, 2009, pp. 353–367].

[Hutter, 1990] Dieter Hutter, 1990. Guiding inductive proofs. In [Stickel, 1990, pp. 147–161].

[Hutter, 1994] Dieter Hutter, 1994. Synthesis of induction orderings for existence proofs. In [Bundy, 1994, pp. 29–41].

[Hutter & Bundy, 1999] Dieter Hutter and Alan Bundy, 1999. The design of the CADE-16 Inductive Theorem Prover Contest. In [Ganzinger, 1999, pp. 374–377].

[Hutter & Sengler, 1996] Dieter Hutter and Claus Sengler, 1996. INKA: the next generation. In [McRobbie & Slaney, 1996, pp. 288–292].

[Hutter & Stephan, 2005] Dieter Hutter and Werner Stephan, editors, 2005. *Mechanizing Mathematical Reasoning: Essays in Honor of Jörg Siekmann on the Occasion of His 60ᵗʰ Birthday*. Number 2605 in Lecture Notes in Artificial Intelligence. Springer.

[IEEE WESTON, 1970] IEEE WESTON, 1970. *Proc. IEEE WESCON, Aug. 1970*. IEEE

Press, originally published by TRW software series, TRW–SS–70–01.

[Ireland & Bundy, 1994] Andrew Ireland and Alan Bundy, 1994. *Productive Use of Failure in Inductive Proof*. DAI Research Paper No. 716, Dept. Artificial Intelligence, Univ. Edinburgh. Also in: J. Automated Reasoning 16:79–111, 1996, Kluwer (Springer Science+Business Media).

[Jamnik &al., 2003] Mateja Jamnik, Manfred Kerber, Martin Pollet, and Christoph Benzmüller, 2003. Automatic learning of proof methods in proof planning. *Logic J. of the IGPL*, 11:647–673.

[Jouannaud & Kounalis, 1986] Jean-Pierre Jouannaud and Emmanuël Kounalis, 1986. Automatic proofs by induction in equational theories without constructors. In [LICS, 1986, pp. 358–366]. Also in Information and Computation 82:1–33, 1989, Academic Press (Elsevier), 1989.

[Kaplan & Jouannaud, 1988] Stéphane Kaplan and Jean-Pierre Jouannaud, editors, 1988. *1st Int. Workshop on Conditional Term Rewriting Systems (CTRS), Orsay (France), 1987*, number 308 in Lecture Notes in Computer Science.

[Kaplan & Okada, 1991] Stéphane Kaplan and Mitsuhiro Okada, editors, 1991. *2nd Int. Workshop on Conditional Term Rewriting Systems (CTRS), Montreal, 1990*, number 516 in Lecture Notes in Computer Science.

[Kapur, 1992] Deepak Kapur, editor, 1992. *11th Int. Conf. on Automated Deduction (CADE), Saratoga Springs (NY), 1992*, number 607 in Lecture Notes in Artificial Intelligence. Springer.

[Kapur & Musser, 1986] Deepak Kapur and David R. Musser, 1986. Inductive reasoning with incomplete specifications. In [LICS, 1986, pp. 367–377].

[Kapur & Musser, 1987] Deepak Kapur and David R. Musser, 1987. Proof by consistency. *Artificial Intelligence*, 31:125–157.

[Kapur & Subramaniam, 1996] Deepak Kapur and Mahadevan Subramaniam, 1996. Automating induction over mutually recursive functions. In [Wirsing & Nivat, 1996, pp. 117–131].

[Kapur & Zhang, 1989] Deepak Kapur and Hantao Zhang, 1989. An overview of Rewrite Rule Laboratory (RRL). In [Dershowitz, 1989, pp. 559–563]. Journal version is [Kapur & Zhang, 1995].

[Kapur & Zhang, 1995] Deepak Kapur and Hantao Zhang, 1995. An overview of Rewrite Rule Laboratory (RRL). *Computers and Mathematics with Applications*, 29(2):91–114.

[Katz, 1998] Victor J. Katz, 1998. *A History of Mathematics: An Introduction*. Addison-Wesley, Reading (MA). 2nd edn..

[Kaufmann &al., 2000a] Matt Kaufmann, Panagiotis Manolios, and J Strother Moore, editors, 2000a. *Computer-Aided Reasoning: ACL2 Case Studies*. Number 4 in Advances in Formal Methods. Kluwer (Springer Science+Business Media). With a foreword from the series editor Mike Hinchey.

[Kaufmann &al., 2000b] Matt Kaufmann, Panagiotis Manolios, and J Strother Moore, 2000b. *Computer-Aided Reasoning: An Approach*. Number 3 in Advances in Formal Methods. Kluwer (Springer Science+Business Media). With a foreword from the series editor Mike Hinchey.

[Kleene, 1952] Stephen C. Kleene, 1952. *Introduction to Metamathematics*. D. Van Nostrand, Princeton (NJ); North-Holland (Elsevier), Amsterdam.

[Knuth & Bendix, 1970] Donald E Knuth and Peter B. Bendix, 1970. Simple word problems in universal algebra. In [Leech, 1970, pp. 263–297].

[Kodratoff, 1988] Yves Kodratoff, editor, 1988. *Proc. 8th European Conf. on Artificial Intelligence (ECAI)*. Pitman Publ., London.

[Kott, 1986] Laurent Kott, editor, 1986. *13th Int. Colloquium on Automata, Languages and Programming (ICALP), Rennes (France)*, number 226 in Lecture Notes in Computer Science. Springer.

[Kowalski, 1974] Robert A. Kowalski, 1974. Predicate logic as a programming language. In [Rosenfeld, 1974, pp. 569–574].

[Kowalski, 1988] Robert A. Kowalski, 1988. The early years of logic programming. *Comm. ACM*, 31:38–43.

[Krazer, 1905] A. Krazer, editor, 1905. *Verhandlungen des Dritten Internationalen Mathematiker-Kongresses, Heidelberg, Aug. 8–13, 1904*. Verlag von B. G. Teubner, Leipzig.

[Kreisel, 1965] Georg Kreisel, 1965. Mathematical logic. In [Saaty, 1965, Vol. III, pp. 95–195].

[Küchlin, 1989] Wolfgang Küchlin, 1989. Inductive completion by ground proof transformation. In [Aït-Kaci & Nivat, 1989, Vol. 2, pp. 211–244].

[Kühler, 1991] Ulrich Kühler, 1991. Ein funktionaler und struktureller Vergleich verschiedener Induktionsbeweiser. (English translation of title: "A functional and structural comparsion of several inductive theorem-proving systems" (INKA, LP (Larch Prover), NQTHM, RRL, UNICOM)). vi+143 pp., Diplomarbeit (Master's thesis), FB Informatik, Univ. Kaiserslautern.

[Kühler, 2000] Ulrich Kühler, 2000. *A Tactic-Based Inductive Theorem Prover for Data Types with Partial Operations*. Infix, Akademische Verlagsgesellschaft Aka GmbH, Sankt Augustin, Berlin. PhD thesis, Univ. Kaiserslautern, ISBN 1586031287, http://wirth.bplaced.net/p/kuehlerdiss.

[Kühler & Wirth, 1996] Ulrich Kühler and Claus-Peter Wirth, 1996. *Conditional Equational Specifications of Data Types with Partial Operations for Inductive Theorem Proving*. SEKI-Report SR–1996–11 (ISSN 1437–4447). SEKI Publications, FB Informatik, Univ. Kaiserslautern. 24 pp., http://wirth.bplaced.net/p/rta97. Short version is [Kühler & Wirth, 1997].

[Kühler & Wirth, 1997] Ulrich Kühler and Claus-Peter Wirth, 1997. Conditional equational specifications of data types with partial operations for inductive theorem proving. In [Comon, 1997, pp. 38–52]. Extended version is [Kühler & Wirth, 1996].

[Lambert, 1764] Johann Heinrich Lambert, 1764. *Neues Organon oder Gedanken über die Erforschung und Bezeichnung des Wahren und dessen Unterscheidung von Irrthum und Schein*. Johann Wendler, Leipzig. Vol. I (Dianoiologie oder die Lehre von den Gesetzen des Denkens, Alethiologie oder Lehre von der Wahrheit) (http://books.google.de/books/about/Neues_Organon_oder_Gedanken_Uber_die_Erf.html?id=ViS3XCuJEw8C) & Vol. II (Semiotik oder Lehre von der Bezeichnung der Gedanken und Dinge, Phänomenologie oder Lehre von dem Schein) (http://books.google.de/books/about/Neues_Organon_oder_Gedanken_%C3%BCber_die_Er.html?id=X8UAAAAAcAAj). Facsimile reprint by Georg Olms Verlag, Hildesheim (Germany), 1965, with a German introduction by Hans Werner Arndt.

[Lankford, 1980] Dallas S. Lankford, 1980. Some remarks on inductionless induction. Memo MTP-11, Math. Dept., Louisiana Tech. Univ., Ruston (LA).

[Lankford, 1981] Dallas S. Lankford, 1981. A simple explanation of inductionless induction.

Memo MTP-14, Math. Dept., Louisiana Tech. Univ., Ruston (LA).

[Lassez & Plotkin, 1991] Jean-Louis Lassez and Gordon D. Plotkin, editors, 1991. *Computational Logic — Essays in Honor of J. Alan Robinson*. MIT Press.

[Leech, 1970] John Leech, editor, 1970. *Computational Word Problems in Abstract Algebra — Proc. of a Conf. held at Oxford, under the auspices of the Science Research Council, Atlas Computer Laboratory, 29th Aug. to 2nd Sept. 1967*. Pergamon Press, Oxford. With a foreword by J. Howlett.

[Leeuwen, 1990] Jan van Leeuwen, editor, 1990. *Handbook of Theoretical Computer Sci.*. MIT Press.

[LICS, 1986] LICS, 1986. *Proc. 1st Annual IEEE Symposium on Logic In Computer Sci. (LICS), Cambridge (MA), 1986*. IEEE Press. http://lii.rwth-aachen.de/lics/archive/1986.

[LICS, 1988] LICS, 1988. *Proc. 3rd Annual IEEE Symposium on Logic In Computer Sci. (LICS), Edinburgh, 1988*. IEEE Press. http://lii.rwth-aachen.de/lics/archive/1988.

[LICS, 2007] LICS, 2007. *Proc. 22nd Annual IEEE Symposium on Logic In Computer Sci. (LICS), Wrocław (i.e. Breslau, Silesia), 2007*. IEEE Press. http://lii.rwth-aachen.de/lics/archive/2007.

[Löchner, 2006] Bernd Löchner, 2006. Things to know when implementing LPO. *Int. J. Artificial Intelligence Tools*, 15:53–79.

[Lusk & Overbeek, 1988] Ewing Lusk and Ross Overbeek, editors, 1988. *9th Int. Conf. on Automated Deduction (CADE), Argonne National Laboratory (IL), 1988*, number 310 in Lecture Notes in Artificial Intelligence. Springer.

[Mahoney, 1994] Michael Sean Mahoney, 1994. *The Mathematical Career of Pierre de Fermat 1601–1665*. Princeton Univ. Press. 2nd rev. edn. (1st edn. 1973).

[Marchisotto & Smith, 2007] Elena Anne Marchisotto and James T. Smith, 2007. *The Legacy of Mario Pieri in Geometry and Arithmetic*. Birkhäuser (Springer), Basel.

[Margaria, 1995] Tiziana Margaria, editor, 1995. *Kolloquium Programmiersprachen und Grundlagen der Programmierung*. Tech. Report MIP–9519, Univ. Passau.

[McCarthy &al., 1965] John McCarthy, Paul W. Abrahams, D. J. Edwards, T. P. Hart, and M. I. Levin, 1965. *LISP 1.5 Programmer's Manual*. MIT Press.

[McRobbie & Slaney, 1996] Michael A. McRobbie and John K. Slaney, editors, 1996. *13th Int. Conf. on Automated Deduction (CADE), New Brunswick (NJ), 1996*, number 1104 in Lecture Notes in Artificial Intelligence. Springer.

[Melis &al., 2008] Erica Melis, Andreas Meier, and Jörg Siekmann, 2008. Proof planning with multiple strategies. *Artificial Intelligence*, 172:656–684. Received May 2, 2006. Published online Nov. 22, 2007. http://dx.doi.org/10.1016/j.artint.2007.11.004.

[Meltzer, 1975] Bernard Meltzer, 1975. Department of A.I. – Univ. of Edinburgh. *ACM SIGART Bulletin*, 50:5.

[Meltzer & Michie, 1972] Bernard Meltzer and Donald Michie, editors, 1972. *Proceedings of the 7th Annual Machine Intelligence Workshop (Machine Intelligence 7), Edinburgh, 1971*. Univ. Edinburgh Press. http://aitopics.org/sites/default/files/classic/Machine%20Intelligence%203/Machine%20Intelligence%20v3.pdf.

[Michie, 1968] Donald Michie, editor, 1968. *Proceedings of the 3rd Annual Machine Intelligence Workshop (Machine Intelligence 3), Edinburgh, 1967*. Univ. Edinburgh Press. http://aitopics.org/sites/default/files/classic/Machine%

20Intelligence%203/Machine%20Intelligence%20v3.pdf.

[Milner, 1972] Robin Milner, 1972. Logic for computable functions — description of a machine interpretation. Technical Report Memo AIM–169, STAN–CS–72–288, Dept. Computer Sci., Stanford University. ftp://reports.stanford.edu/pub/cstr/reports/cs/tr/72/288/CS-TR-72-288.pdf.

[Moore, 1973] J Strother Moore, 1973. *Computational Logic: Structure Sharing and Proof of Program Properties*. PhD thesis, Dept. Artificial Intelligence, Univ. Edinburgh. http://hdl.handle.net/1842/2245.

[Moore, 1975a] J Strother Moore, 1975a. Introducing iteration into the PURE LISP THEOREM PROVER. Technical Report CSL 74–3, Xerox, Palo Alto Research Center, 3333 Coyote Hill Rd., Palo Alto (CA). ii+37 pp., Received Dec. 1974, rev. March 1975. Short version is [Moore, 1975b].

[Moore, 1975b] J Strother Moore, 1975b. Introducing iteration into the PURE LISP THEOREM PROVER. *IEEE Transactions on Software Engineering*, 1:328–338. http://doi.ieeecomputersociety.org/10.1109/TSE.1975.6312857. Long version is [Moore, 1975a].

[Moore, 1979] J Strother Moore, 1979. A mechanical proof of the termination of Takeuti's function. *Information Processing Letters*, 9:176–181. Received July 13, 1979. Rev. Sept. 5, 1979. http://dx.doi.org/10.1016/0020-0190(79)90063-2.

[Moore, 1981] J Strother Moore, 1981. Text editing primitives — the TXDT package. Technical Report CSL 81–2, Xerox, Palo Alto Research Center, 3333 Coyote Hill Rd., Palo Alto (CA).

[Moore, 1989a] J Strother Moore, 1989a. A mechanically verified language implementation. *J. Automated Reasoning*, 5:461–492.

[Moore, 1989b] J Strother Moore, 1989b. System verification. *J. Automated Reasoning*, 5:409–410.

[Moore &al., 1998] J Strother Moore, Thomas Lynch, and Matt Kaufmann, 1998. A mechanically checked proof of the correctness of the kernel of the AMD5K86 floating point division algorithm. *IEEE Transactions on Computers*, 47:913–926.

[Moskewicz &al., 2001] Matthew W. Moskewicz, Conor F. Madigan, Ying Zhao, Lintao Zhang, and Sharad Malik, 2001. CHAFF: Engineering an efficient SAT solver. In [DAC, 2001, pp. 530–535].

[Musser, 1980] David R. Musser, 1980. On proving inductive properties of abstract data types. In [Abrahams &al., 1980, pp. 154–162]. http://dl.acm.org/citation.cfm?id=567461.

[Nilsson, 1973] Nils J. Nilsson, editor, 1973. *Proc. 3rd Int. Joint Conf. on Artificial Intelligence (IJCAI), Stanford (CA)*. Stanford Research Institute, Publications Dept., Stanford (CA). http://ijcai.org/Past%20Proceedings/IJCAI-73/CONTENT/content.htm.

[Odifreddi, 1990] Piergiorgio Odifreddi, editor, 1990. *Logic and Computer Science*. Academic Press (Elsevier).

[Padawitz, 1996] Peter Padawitz, 1996. Inductive theorem proving for design specifications. *J. Symbolic Computation*, 21:41–99.

[Padoa, 1913] Alessandro Padoa, 1913. La valeur et les rôles du principe d'induction mathématique. In [Hobson & Love, 1913, pp. 471–479].

[Pascal, 1954] Blaise Pascal, 1954. *Œuvres Complètes*. Gallimard, Paris. Ed. by Jacques Chevalier.

[Paulson, 1990] Lawrence C. Paulson, 1990. ISABELLE: The next 700 theorem provers. In [Odifreddi, 1990, pp. 361–386].

[Paulson, 1996] Lawrence C. Paulson, 1996. ML *for the Working Programmer.* Cambridge Univ. Press. 2nd edn. (1st edn. 1991).

[Peano, 1889] Guiseppe Peano, 1889. *Arithmetices principia – nova methodo exposita.* Fratelli Bocca, Torino (i.e. Turin, Italy). Title page actually says: Ioseph Peano, Fratres Bocca, Augustae Taurinorum.

[Péter, 1932] Rózsa Péter, 1932. Rekursive Funktionen. In [Saxer, 1932, Vol. II, p. 336]. Actually published under the name Rózsa Politzer.

[Péter, 1935] Rózsa Péter, 1935. Über den Zusammenhang der verschiedenen Begriffe der rekursiven Funktion. *Mathematische Annalen*, 110:612–632. Received Sept. 21, 1934.

[Péter, 1951] Rózsa Péter, 1951. *Rekursive Funktionen.* Akad. Kiadó, Budapest. 1st edn..

[Péter, 1957] Rózsa Péter, 1957. *Rekursive Funktionen.* Akad. Kiadó, Budapest. 2nd extd. edn. (1st edn. 1951). English translation is [Péter, 1967].

[Péter, 1967] Rózsa Péter, 1967. *Recursive Functions.* Akad. Kiadó, Budapest; joint edn. with Academic Press (Elsevier). 3rd rev. edn., 1st edn. in English, translated form the German [Péter, 1957] by István Földes.

[Pieri, 1908] Mario Pieri, 1908. Sopra gli assiomi aritmetici. *Il Bollettino delle seduta della Accademia Gioenia di Scienze Naturali in Catania*, Series 2, 1–2:26–30. Written Dec. 1907. Received Jan. 8, 1908. English translation *On the Axioms of Arithmetic* in [Marchisotto & Smith, 2007, § 4.2, pp. 308–313].

[Plotkin &al., 2000] Gordon D. Plotkin, Colin Stirling, and Mads Tofte, editors, 2000. *Proof, Language, and Interaction, Essays in Honour of Robin Milner.* MIT Press.

[Presburger, 1930] Mojżesz Presburger, 1930. Über die Vollständigkeit eines gewissen Systems der Arithmetik ganzer Zahlen, in welchem die Addition als einzige Operation hervortritt. In *Sprawozdanie z I Kongresu metematyków krajów słowianskich, Warszawa 1929 (Comptes-rendus du 1re Congrès des Mathématiciens des Pays Slaves, , Varsovie 1929)*, pages 92–101+395. Remarks and English translation in [Stansifer, 1984].

[Protzen, 1994] Martin Protzen, 1994. Lazy generation of induction hypotheses. In [Bundy, 1994, pp. 42–56].

[Protzen, 1995] Martin Protzen, 1995. *Lazy Generation of Induction Hypotheses and Patching Faulty Conjectures.* Infix, Akademische Verlagsgesellschaft Aka GmbH, Sankt Augustin, Berlin. PhD thesis.

[Protzen, 1996] Martin Protzen, 1996. Patching faulty conjectures. In [McRobbie & Slaney, 1996, pp. 77–91].

[Rabinovitch, 1970] Nachum L. Rabinovitch, 1970. Rabbi Levi ben Gerson and the origins of mathematical induction. *Archive for History of Exact Sciences*, 6:237–248. Received Jan. 12, 1970.

[Reddy, 1977] Ray Reddy, editor, 1977. *Proc. 5th Int. Joint Conf. on Artificial Intelligence (IJCAI), Cambridge (MA).* Dept. of Computer Sci., Carnegie Mellon Univ., Cambridge (MA). http://ijcai.org/Past%20Proceedings.

[Reddy, 1990] Uday S. Reddy, 1990. Term rewriting induction. [Stickel, 1990, pp. 162–177].

[Riazanov & Voronkov, 2001] Alexander Riazanov and Andrei Voronkov, 2001. Vampire 1.1 (system description). In [Gore &al., 2001, pp. 376–380].

[Robinson & Voronkov, 2001] J. Alan Robinson and Andrei Voronkow, editors, 2001. *Hand-*

book of Automated Reasoning. Elsevier.

[Rosenfeld, 1974] Jack L. Rosenfeld, editor, 1974. *Proc. of the Congress of the Int. Federation for Information Processing (IFIP), Stockholm (Sweden), Aug. 5–10, 1974.* North-Holland (Elsevier).

[Royce, 1970] Winsten W. Royce, 1970. Managing the development of large software systems. In [IEEE WESTON, 1970, pp. 1–9].

[Rubin & Rubin, 1985] Herman Rubin and Jean E. Rubin, 1985. *Equivalents of the Axiom of Choice.* North-Holland (Elsevier). 2nd rev. edn. (1st edn. 1963).

[Rusinowitch & Remy, 1993] Michaël Rusinowitch and Jean-Luc Remy, editors, 1993. *3rd Int. Workshop on Conditional Term Rewriting Systems (CTRS), Pont-à-Mousson (France), 1992*, number 656 in Lecture Notes in Computer Science.

[Russinoff, 1998] David M. Russinoff, 1998. A mechanically checked proof of IEEE compliance of a register-transfer-level specification of the AMD-K7 floating-point multiplication, division, and square root instructions. *London Mathematical Society Journal of Computation and Mathematics*, 1:148–200.

[Saaty, 1965] T. L. Saaty, editor, 1965. *Lectures on Modern Mathematics.* John Wiley & Sons, New York.

[Saxer, 1932] Walter Saxer, editor, 1932. *Verhandlungen des Internationalen Mathematiker-Kongresses, Zürich, 1932.* Verlag Orell Füssli, Zürich.

[Schmidt-Samoa, 2006a] Tobias Schmidt-Samoa, 2006a. An even closer integration of linear arithmetic into inductive theorem proving. *Electronic Notes in Theoretical Computer Sci.*, 151:3–20. http://wirth.bplaced.net/p/evencloser, http://dx.doi.org/10.1016/j.entcs.2005.11.020.

[Schmidt-Samoa, 2006b] Tobias Schmidt-Samoa, 2006b. *Flexible Heuristic Control for Combining Automation and User-Interaction in Inductive Theorem Proving.* PhD thesis, Univ. Kaiserslautern. http://wirth.bplaced.net/p/samoadiss.

[Schmidt-Samoa, 2006c] Tobias Schmidt-Samoa, 2006c. Flexible heuristics for simplification with conditional lemmas by marking formulas as forbidden, mandatory, obligatory, and generous. *J. Applied Non-Classical Logics*, 16:209–239. http://dx.doi.org/10.3166/jancl.16.208-239.

[Scott, 1993] Dana S. Scott, 1993. A type-theoretical alternative to ISWIM, CUCH, OWHY. *Theoretical Computer Sci.*, 121:411–440. Annotated version of a manuscript from the year 1969. www.cs.cmu.edu/~kw/scans/scott93tcs.pdf.

[Shankar, 1986] Natarajan Shankar, 1986. *Proof-checking Metamathematics.* PhD thesis, The University of Texas at Austin. Thoroughly rev. version is [Shankar, 1994].

[Shankar, 1988] Natarajan Shankar, 1988. A mechanical proof of the Church–Rosser theorem. *J. of the ACM*, 35:475–522. Received May 1985, rev. Aug. 1987. See also Chapter 6 in [Shankar, 1994].

[Shankar, 1994] Natarajan Shankar, 1994. *Metamathematics, Machines, and Gödel's Proof.* Cambridge Univ. Press. Originally published as [Shankar, 1986]. Paperback reprint 1997.

[Shoenfield, 1967] Joseph R. Shoenfield, 1967. *Mathematical Logic.* Addison–Wesley, Reading (MA). 1st edn. 1967; 2nd edn. 1973; 3rd edn. 2001, facsimile of 2nd edn., A K Peters, Natick (MA), copyright 1967 by the Association for Symbolic Logic.

[Siekmann, 1986] Jörg Siekmann, editor, 1986. *8th Int. Conf. on Automated Deduction (CADE), Oxford, 1986*, number 230 in Lecture Notes in Artificial Intelligence. Springer.

[Sridharan, 1989] N. S. Sridharan, editor, 1989. *Proc. 11th Int. Joint Conf. on Artificial Intelligence (IJCAI), Detroit (MI)*. Morgan Kaufmann (Elsevier), Los Altos (CA). http://ijcai.org/Past%20Proceedings.

[Stansifer, 1984] Ryan Stansifer, 1984. Presburger's Article on Integer Arithmetic: Remarks and Translation. Technical Report TR 84–639, Dept. of Computer Sci., Cornell Univ., Ithaca (NY). http://hdl.handle.net/1813/6478.

[Steele, 1990] Guy L. Steele Jr., 1990. COMMON LISP — *The Language*. Digital Press (Elsevier). 2nd edn. (1st edn. 1984).

[Steinbach, 1988] Joachim Steinbach, 1988. *Term Orderings With Status*. SEKI-Report SR-88-12 (ISSN 1437–4447). SEKI Publications, FB Informatik, Univ. Kaiserslautern. 57 pp., http://wirth.bplaced.net/SEKI/welcome.html#SR-88-12.

[Steinbach, 1995] Joachim Steinbach, 1995. Simplification orderings — history of results. *Fundamenta Informaticae*, 24:47–87.

[Stevens, 1988] Andrew Stevens, 1988. *A Rational Reconstruction of Boyer and Moore's Technique for Constructing Induction Formulas*. DAI Research Paper No. 360, Dept. Artificial Intelligence, Univ. Edinburgh. Also in [Kodratoff, 1988, pp. 565–570].

[Stickel, 1990] Mark E. Stickel, editor, 1990. *10th Int. Conf. on Automated Deduction (CADE), Kaiserslautern (Germany), 1990*, number 449 in Lecture Notes in Artificial Intelligence. Springer.

[Stoyan, 1985] Herbert Stoyan, editor, 1985. *9th German Workshop on Artificial Intelligence (GWAI), Dassel (Germany), 1985*, number 118 in Informatik-Fachberichte. Springer.

[Toyama, 1988] Yoshihito Toyama, 1988. Commutativity of term rewriting systems. In [Fuchi & Kott, 1988, pp. 393–407]. Also in [Toyama, 1990].

[Toyama, 1990] Yoshihito Toyama, 1990. *Term Rewriting Systems and the Church–Rosser Property*. PhD thesis, Tohoku Univ. / Nippon Telegraph and Telephone Corporation.

[Unguru, 1991] Sabetai Unguru, 1991. Greek mathematics and mathematical induction. *Physis*, XXVIII(2):273–289.

[Verma, 2005?] Shamit Verma, 2005? Interview with Charles Simonyi. WWW only: http://www.shamit.org/charles_simonyi.htm.

[Voicu & Li, 2009] Răzvan Voicu and Mengran Li, 2009. *Descente Infinie* proofs in COQ. In [Herbelin, 2009, pp. 73–84].

[Voronkov, 1992] Andrei Voronkov, editor, 1992. *Proc. 3rd Int. Conf. on Logic for Programming, Artificial Intelligence, and Reasoning (LPAR)*, number 624 in Lecture Notes in Artificial Intelligence. Springer.

[Voronkov, 2002] Andrei Voronkov, editor, 2002. *18th Int. Conf. on Automated Deduction (CADE), København (Denmark), 2002*, number 2392 in Lecture Notes in Artificial Intelligence. Springer.

[Walther, 1988] Christoph Walther, 1988. Argument-bounded algorithms as a basis for automated termination proofs. In [Lusk & Overbeek, 1988, pp. 601–622].

[Walther, 1992] Christoph Walther, 1992. Computing induction axioms. In [Voronkov, 1992, pp. 381–392].

[Walther, 1993] Christoph Walther, 1993. Combining induction axioms by machine. In [Bajscy, 1993, pp. 95–101].

[Walther, 1994a] Christoph Walther, 1994a. Mathematical induction.

In [Gabbay &al., 1994, pp. 127–228].

[Walther, 1994b] Christoph Walther, 1994b. On proving termination of algorithms by machine. *Artificial Intelligence*, 71:101–157.

[Wirsing & Nivat, 1996] Martin Wirsing and Maurice Nivat, editors, 1996. *Proc. 5th Int. Conf. on Algebraic Methodology and Software Technology (AMAST), München (Germany), 1996*, number 1101 in Lecture Notes in Computer Science. Springer.

[Wirth, 1991] Claus-Peter Wirth, 1991. Inductive theorem proving in theories specified by positive/negative-conditional equations. Diplomarbeit (Master's thesis), FB Informatik, Univ. Kaiserslautern.

[Wirth, 1997] Claus-Peter Wirth, 1997. *Positive/Negative-Conditional Equations: A Constructor-Based Framework for Specification and Inductive Theorem Proving*, volume 31 of *Schriftenreihe Forschungsergebnisse zur Informatik*. Verlag Dr. Kovač, Hamburg. PhD thesis, Univ. Kaiserslautern, ISBN 386064551X, http://wirth.bplaced.net/p/diss.

[Wirth, 2004] Claus-Peter Wirth, 2004. Descente Infinie + Deduction. *Logic J. of the IGPL*, 12:1–96. http://wirth.bplaced.net/p/d.

[Wirth, 2005a] Claus-Peter Wirth, 2005a. History and future of implicit and inductionless induction: Beware the old jade and the zombie! In [Hutter & Stephan, 2005, pp. 192–203], http://wirth.bplaced.net/p/zombie.

[Wirth, 2005b] Claus-Peter Wirth, 2005b. *Syntactic Confluence Criteria for Positive/Negative-Conditional Term Rewriting Systems*. SEKI-Report SR–95–09 (ISSN 1437–4447). SEKI Publications, Univ. Kaiserslautern. Rev. edn. Oct. 2005 (1st edn. 1995), ii+188 pp., http://arxiv.org/abs/0902.3614.

[Wirth, 2009] Claus-Peter Wirth, 2009. Shallow confluence of conditional term rewriting systems. *J. Symbolic Computation*, 44:69–98. http://dx.doi.org/10.1016/j.jsc.2008.05.005.

[Wirth, 2010a] Claus-Peter Wirth, 2010a. *Progress in Computer-Assisted Inductive Theorem Proving by Human-Orientedness and Descente Infinie?* SEKI-Working-Paper SWP–2006–01 (ISSN 1860–5931). SEKI Publications, Saarland Univ. Rev. edn. Dec 2010 (1st edn. 2006), ii+36 pp., http://arxiv.org/abs/0902.3294.

[Wirth, 2010b] Claus-Peter Wirth, 2010b. *A Self-Contained and Easily Accessible Discussion of the Method of Descente Infinie and Fermat's Only Explicitly Known Proof by Descente Infinie*. SEKI-Working-Paper SWP–2006–02 (ISSN 1860–5931). SEKI Publications. Rev. ed. Dec. 2010, ii+36 pp., http://arxiv.org/abs/0902.3623.

[Wirth, 2012a] Claus-Peter Wirth, 2012a. Herbrand's Fundamental Theorem in the eyes of Jean van Heijenoort. *Logica Universalis*, 6:485–520. Received Jan. 12, 2012. Published online June 22, 2012, http://dx.doi.org/10.1007/s11787-012-0056-7.

[Wirth, 2012b] Claus-Peter Wirth, 2012b. lim+, δ^+, and Non-Permutability of β-Steps. *J. Symbolic Computation*, 47:1109–1135. Received Jan. 18, 2011. Published online July 15, 2011, http://dx.doi.org/10.1016/j.jsc.2011.12.035.

[Wirth, 2012c] Claus-Peter Wirth, 2012c. Human-oriented inductive theorem proving by descente infinie — a manifesto. *Logic J. of the IGPL*, 20:1046–1063. Received July 11, 2011. Published online March 12, 2012, http://dx.doi.org/10.1093/jigpal/jzr048.

[Wirth, 2012d] Claus-Peter Wirth, 2012d. Unpublished Interview of Robert S. Boyer and J Strother Moore at Boyer's place in Austin (TX) on Thursday, Oct. 7.

[Wirth, 2017] Claus-Peter Wirth, 2017. A simplified and improved free-variable framework for Hilbert's epsilon as an operator of indefinite committed choice. *IFCoLog J. of Logics and Their Applications*, 4:435–526. Received Oct. 23, 2015.

[Wirth & Becker, 1995] Claus-Peter Wirth and Klaus Becker, 1995. Abstract notions and inference systems for proofs by mathematical induction.
In [Dershowitz & Lindenstrauss, 1995, pp. 353–373].

[Wirth & Gramlich, 1994a] Claus-Peter Wirth and Bernhard Gramlich, 1994a. A constructor-based approach to positive/negative-conditional equational specifications. *J. Symbolic Computation*, 17:51–90. http://dx.doi.org/10.1006/jsco.1994.1004, http://wirth.bplaced.net/p/jsc94.

[Wirth & Gramlich, 1994b] Claus-Peter Wirth and Bernhard Gramlich, 1994b. On notions of inductive validity for first-order equational clauses. In [Bundy, 1994, pp. 162–176], http://wirth.bplaced.net/p/cade94.

[Wirth &al., 1993] Claus-Peter Wirth, Bernhard Gramlich, Ulrich Kühler, and Horst Prote, 1993. *Constructor-Based Inductive Validity in Positive/Negative-Conditional Equational Specifications*. SEKI-Report SR–93–05 (SFB) (ISSN 1437–4447). SEKI Publications, FB Informatik, Univ. Kaiserslautern. iv+58 pp., http://wirth.bplaced.net/SEKI/welcome.html#SR-93-05. Rev. extd. edn. of 1st part is [Wirth & Gramlich, 1994a], rev. edn. of 2nd part is [Wirth & Gramlich, 1994b].

[Wirth & Kühler, 1995] Claus-Peter Wirth and Ulrich Kühler, 1995. *Inductive Theorem Proving in Theories Specified by Positive/Negative-Conditional Equations*. SEKI-Report SR–95–15 (ISSN 1437–4447). SEKI Publications, Univ. Kaiserslautern. iv+126 pp..

[Wirth &al., 2009] Claus-Peter Wirth, Jörg Siekmann, Christoph Benzmüller, and Serge Autexier, 2009. Jacques Herbrand: Life, logic, and automated deduction. In [Gabbay & Woods, 2004ff., Vol. 5: Logic from Russell to Church, pp. 195–254].

[Wirth, 1971] Niklaus Wirth, 1971. The programming language PASCAL. *Acta Informatica*, 1:35–63.

[Wolff, 1728] Christian Wolff, 1728. *Philosophia rationalis sive Logica, methodo scientifica pertractata et ad usum scientiarium atque vitae aptata*. Rengerische Buchhandlung, Frankfurt am Main & Leipzig. 1st edn..

[Wolff, 1740] Christian Wolff, 1740. *Philosophia rationalis sive Logica, methodo scientifica pertractata et ad usum scientiarium atque vitae aptata*. Rengerische Buchhandlung, Frankfurt am Main & Leipzig. 3rd extd. edn. of [Wolff, 1728]. Facsimile reprint by Georg Olms Verlag, Hildesheim (Germany), 1983, with a French introduction by Jean École.

[Yeh & Ramamoorthy, 1976] Raymond T. Yeh and C. V. Ramamoorthy, editors, 1976. *Proc. 2nd Int. Conf. on Software Engineering, San Francisco (CA), Oct. 13–15, 1976*. IEEE Computer Sci. Press, Los Alamitos (CA). http://dl.acm.org/citation.cfm?id=800253.

[Young, 1989] William D. Young, 1989. A mechanically verified code generator. *J. Automated Reasoning*, 5:493–518.

[Zhang &al., 1988] Hantao Zhang, Deepak Kapur, and Mukkai S. Krishnamoorthy, 1988. A mechanizable induction principle for equational specifications.
In [Lusk & Overbeek, 1988, pp. 162–181].

[Zygmunt, 1991] Jan Zygmunt, 1991. Mojżesz Presburger: Life and work. *History and Philosophy of Logic*, 12:211–223.

Index

77-editor, 1508

accessor functions, 1567
Acerbi, Fabio (*1965), 1514, 1605, 1606
Ackermann function, 1519–1521, 1546, 1565, 1582
Ackermann, Wilhelm (1896–1962), 1520, 1526, 1606
ACL2, 1510, 1512, 1513, 1553, 1565, 1574, 1589, 1590, 1592–1595, 1603, 1620
Aristotle (384–322 B.C.), 1514
Aubin, Raymond, 1536, 1597, 1599, 1606
Autexier, Serge (*1971), 1528, 1595, 1606, 1628
Axiom of Choice, see choice, Axiom of Choice
Axiom of Structural Induction, see induction, structural

Barner, Klaus, 1515, 1605, 1607
BAROQUE, 1509
Basin, David, 1598, 1607, 1612
Bauer-Mengelberg, Stefan (1927–1996), 1617
Benzmüller, Christoph (*1968), 1506, 1607, 1620, 1628
Bernays, Paul (1888–1977), 1519, 1520, 1525, 1526, 1607, 1616, 1618, 1619
Bledsoe, W. W. (1921–1995), 1507, 1508, 1553–1555, 1608, 1610
Bourbaki, Nicolas (pseudonym), 1516, 1517, 1608, 1609, 1612
Boyer, Robert S. (*1946), 1506–1513, 1524, 1531–1535, 1537, 1540–1544, 1547, 1548, 1553, 1554, 1556–1567, 1569–1571, 1573, 1574, 1576, 1577, 1579, 1580, 1582, 1585–1587, 1591–1595, 1598, 1599, 1603, 1605, 1608–1611, 1627
Boyer–Moore fast string searching algorithm, 1610
Boyer–Moore machines, 1512
Boyer–Moore theorem provers, 1510, 1512, 1513, 1532–1535, 1537, 1540–1544, 1554, 1556, 1558, 1559, 1561–1565, 1571, 1573, 1574, 1592, 1594, 1595

Boyer–Moore waterfall, 1506, 1507, 1531–1535, 1554, 1557, 1558, 1560, 1564, 1570, 1587
Bundy, Alan (*1947), 1507, 1509, 1510, 1513, 1588, 1595–1598, 1605, 1611, 1612, 1619, 1620, 1624, 1628
Burstall, Rod M. (*1934), 1507–1509, 1553, 1555, 1612

changeable positions, 1547–1551, 1583
choice
    Axiom of Choice, 1516, 1518
    Principle of Dependent Choice, 1516
Church, Alonzo (1903–1995), 1608, 1612
Church–Rosser property, 1533, 1539
Church–Rosser Theorem, 1533
COMMON LISP, 1510, 1512, 1592, 1603, 1626
confluence, 1525, 1533, 1539–1544, 1556, 1602
consistency, 1526, 1538, 1540, 1542, 1543, 1599, 1602
constructor function symbols, 1518, 1543, 1560, 1580
constructor style, 1519, 1524, 1535, 1540, 1562, 1576, 1577
constructor substitutions, 1534, 1535, 1548, 1549, 1551
constructor variables, 1512, 1541–1543, 1547, 1556, 1601
COQ, 1594, 1616, 1626
cross-fertilization, 1506, 1533, 1560–1561, 1576, 1579

Dawson, John W., Jr. (*1944), 1615
Dedekind, Richard (1831–1916), 1519, 1613
descente infinie, 1514, 1515, 1526–1529, 1532, 1533, 1538, 1581, 1587, 1598, 1600, 1601, 1603, 1604, 1626, 1627
destructor elimination, 1506, 1559, 1574–1579, 1587
destructor style, 1519, 1524, 1540, 1541, 1544, 1546, 1562, 1563, 1572, 1574–1577, 1582–1584

elimination of irrelevance, 1506, 1535, 1580–1581

Euclid, 1514, 1613

Feferman, Sol(omon) (1928–2016), 1615
Fermat, Pierre (160?–1665), 1515, 1526, 1527, 1538, 1607, 1615, 1622
Fries, Jakob Friedrich (1773–1843), 1519, 1614

Gabbay, Dov (*1945), 1614, 1618, 1619, 1627, 1628
Gabbay, Michael, 1618, 1619
generalization, 1506, 1535–1537, 1561, 1579
Gentzen, Gerhard (1909–1945), 1525, 1538, 1614
Gerson, Levi ben (1288–1344), 1515, 1624
Gödel, Kurt (1906–1978), 1606, 1615
Goldfarb, Warren (*1949), 1615
Goldstein, Catherine (*1958), 1515, 1605, 1615
Goodstein, R. L. (1912–1985), 1509, 1615
Gordon, Mike J. C. (*1948), 1507, 1509, 1511, 1615
Gramlich, Bernhard (1959–2014), 1512, 1525, 1542, 1543, 1599–1602, 1605, 1611, 1615, 1616, 1628
ground terms, 1539

Harmelen, Frank van (*1960), 1598, 1612
HASKELL, 1511, 1619
Hayes, Pat(rick) J. (*1944), 1507, 1509, 1511
Heijenoort, Jean van (1912–1986), 1519, 1607, 1615–1617, 1627
Herbrand's Fundamental Theorem, 1627
Herbrand, Jacques (1908–1938), 1628
Hilbert, David (1862–1943), 1519, 1520, 1525, 1526, 1616–1619
Hillenbrand, Thomas (*1970), 1505, 1611, 1619
Hippasus of Metapontum (ca. 550 B.C.), 1514, 1614
hitting ratio, 1550, 1551, 1583–1585
HOL, 1594, 1615
Hope Park, 1507, 1508
Howard, Paul (*1943), 1516, 1619
Hunt, Warren A., 1591–1593, 1605, 1608, 1611, 1619
Hutter, Dieter (*1959), 1595, 1598, 1605, 1606, 1608, 1612, 1619, 1627

Huygens, Christiaan (1629–1695), 1527
induction
  complete, 1517
  course-of-values, 1517
  descente infinie, see descente infinie
  explicit, 1530–1539, 1544, 1547–1550, 1552–1601, 1603, 1604
  implicit, 1597–1602
  inductionless, 1600
  lazy, 1549, 1595, 1600, 1601, 1604
  mathematical, 1519
  Noetherian, 1516–1519, 1526–1531, 1534
  structural, 1514, 1515, 1517–1519, 1521, 1525, 1526, 1532, 1533, 1535, 1553, 1556, 1564, 1596
induction rule, 1530–1536, 1544, 1548, 1550, 1558, 1559, 1562–1564, 1576, 1581–1583, 1600
induction schemes, 1548–1551, 1562, 1567, 1576, 1580, 1583–1590, 1603, 1604
induction templates, 1544–1551, 1562, 1573, 1582–1584, 1587, 1589, 1590, 1592, 1603
induction variables, 1534, 1535, 1549, 1550, 1563, 1584–1586, 1588, 1603
Induktion
  vollständige, 1519
INKA, 1595, 1601, 1606, 1619, 1621
Ireland, Andrew, 1596, 1598, 1612, 1620
Isabelle, 1596, 1613, 1624
Isabelle/HOL, 1594
ISAPLANNER, 1594, 1596, 1613

Kant, Immanuel (1724–1804), 1514
Kaufmann, Matt (*1952), 1513, 1566, 1570, 1574, 1592, 1593, 1605, 1620, 1623
Kleene, Stephen C. (1909–1994), 1517, 1615, 1620
Kowalski, Robert A. (*1941), 1507, 1509, 1511, 1608, 1615, 1621
Kühler, Ulrich (*1964), 1543, 1595, 1600–1606, 1621, 1628

LCF, 1511, 1615
LEO-II, 1506, 1607
lexicographic combination, 1525
linear arithmetic, 1591, 1604, 1610, 1625

linear resolution, 1508
linear terms, 1519
LISP, 1506, 1508–1513, 1534, 1535, 1537, 1540, 1542–1544, 1553–1567, 1569–1572, 1574–1576, 1579–1581, 1585, 1586, 1588, 1591, 1592, 1603, 1610, 1623, 1626
Löchner, Bernd (*1967), 1505, 1602, 1619, 1622
logic
    Aristotelian, 1519

Maurolico, Francesco (1494–1575), 1515
McCarthy, John (1927–2011), 1508
measured positions, 1525, 1545–1549, 1565, 1573, 1582–1584
Meltzer, Bernard (1916(?)–2008), 1507, 1509, 1610, 1615, 1622
Michie, Donald (1923–2007), 1509, 1610, 1613, 1616, 1622
Microsoft Word, 1508
Milner, Robin (1934–2010), 1507, 1509, 1624
ML, 1511, 1624
Moore, J Strother (*1947), 1506–1513, 1524, 1531–1535, 1537, 1540–1544, 1547, 1548, 1553–1567, 1569–1571, 1573, 1574, 1576, 1577, 1579, 1580, 1582, 1585–1587, 1591–1595, 1598, 1599, 1603, 1608–1611, 1620, 1623, 1627

Newman Lemma, 1533
Newman, Max(well) H. A. (1897–1984), 1533
Noether, Emmy (1882–1935), 1516, 1613
Noetherian induction, *see* induction, Noetherian
normalation, 1557
NQTHM, 1510, 1513, 1565, 1566, 1574, 1589–1592, 1595, 1601, 1611, 1621
NUPRL, 1596, 1612

OYSTER/CLAM, 1595, 1596, 1612

Péter, Rózsa (1905–1977), 1520, 1624
Padoa, Alessandro (1868–1937), 1520, 1623
Parsons, Charles (*1933), 1615
Pascal, Blaise (1623–1662), 1515
Peano axioms, 1520
Peano, Guiseppe (1858–1932), 1520, 1523, 1526, 1534, 1596, 1624

Peckhaus, Volker (*1955), 1618, 1619
Pieri, Mario (1860–1913), 1520, 1523, 1526, 1534, 1570, 1622, 1624
Plato (427–347 B.C.), 1514
POP-2, 1509, 1555, 1556
position sets, 1549, 1550, 1584, 1585
Presburger Arithmetic, *see* linear arithmetic
Presburger, Mojżesz (1904–1943(?)), 1591, 1624, 1626, 1628
PROLOG, 1509, 1612
proof by consistency, 1599, 1601
proof planning, 1597
Protzen, Martin (*1962), 1595, 1601, 1624
PURE LISP THEOREM PROVER, 1506, 1508–1510, 1513, 1534, 1535, 1537, 1542–1544, 1553–1567, 1569–1572, 1575, 1576, 1579–1581, 1585, 1588, 1592, 1623
PVS, 1594

QTHM, 1565, 1566
QUODLIBET, 1558, 1595, 1597, 1600–1604, 1606

recognizer functions, 1567
recursion, 1539
recursion analysis, 1559, 1564, 1565, 1581–1588, 1596, 1598, 1603, 1604
reducibility, 1541
relational descriptions, 1544–1546, 1548, 1549, 1551, 1582–1584
rewrite relation, 1539
ripple analysis, 1596
rippling, 1595–1598
Robinson, J. Alan (1930–2016), 1508, 1612, 1614, 1622, 1624
RRL, 1594, 1597, 1601, 1620, 1621
Rubin, Herman (*1926), 1516, 1518, 1625
Rubin, Jean E. (1926–2002), 1516, 1518, 1619, 1625
Russell's Paradox, 1542

SATALLAX, 1506, 1611
Schmidt-Samoa, Tobias (*1973), 1543, 1558, 1573, 1603–1606, 1625
Scott, Dana S. (*1932), 1508, 1511, 1625
Shankar, Natarajan, 1533, 1592, 1625
shell principle, 1524, 1567–1569

shells, 1513, 1524, 1567–1570, 1580, 1583
Shoenfield, Joseph R. (1927–2000), 1517, 1542, 1625
Sieg, Wilfried, 1615
Siekmann, Jörg (*1941), 1605, 1608, 1618, 1619, 1622, 1625
Simonyi, Charles, 1508, 1626
simplification, 1506, 1557–1559, 1570–1574
Smith, James T., 1520, 1622, 1624
step-case descriptions, 1549–1551, 1584–1587
structural induction, *see* induction, structural

termination, 1515, 1541, 1544–1547
Theorem of Noetherian Induction, *see* induction, Noetherian
THM, 1510, 1513, 1535, 1544, 1553, 1554, 1565–1592, 1604

UNICOM, 1599, 1601, 1615, 1621

WALDMEISTER, 1505, 1611, 1619
Walther, Christoph (*1950), 1510, 1513, 1532, 1544, 1588, 1595, 1608, 1626, 1627
well-foundedness, 1515
Wirth, Claus-Peter (*1963), 1512, 1514–1516, 1520, 1525–1527, 1533, 1537, 1539, 1541–1543, 1546, 1550, 1555, 1556, 1581, 1600–1606, 1609, 1616, 1618, 1619, 1621, 1627, 1628

Received 15 September 2016

www.ingramcontent.com/pod-product-compliance
Lightning Source LLC
Chambersburg PA
CBHW081457040426
42446CB00016B/3278